ASTRIDE TWO WORLDS

TECHNOLOGY AND THE AMERICAN CIVIL WAR

EDITED BY

BARTON C. HACKER

A SMITHSONIAN CONTRIBUTION TO KNOWLEDGE

Smithsonian Institution
Scholarly Press

WASHINGTON, D.C.
2016

Published by
SMITHSONIAN INSTITUTION SCHOLARLY PRESS
P.O. Box 37012, MRC 957
Washington, D.C. 20013-7012
www.scholarlypress.si.edu

Dust jacket images: Left halves of stereographs courtesy of Library of Congress, Prints and Photographs Division (LOC). *Front:* Amassed Federal artillery and troops near Yorktown, Virginia, during the Peninsula Campaign, May 1862. Photographer unknown. LOC, LC-DIG-cwpb-01580. *Back:* Siege of Yorktown, Virginia, May 1862. Union soldiers mounting 13-inch mortars in Federal Battery No. 4 along the front line. Photo attributed to James F. Gibson. LOC, LC-DIG-stereo-1s02545.

Library of Congress Cataloging-in-Publication Data
Names: Hacker, Barton C., 1935– editor, author.
Title: Astride two worlds : technology and the American Civil War / edited by Barton C. Hacker.
Description: Washington, D.C. : Smithsonian Institution Scholarly Press, 2016. | "A Smithsonian contribution to knowledge." | Includes bibliographical references and index.
Identifiers: LCCN 2015048163 | ISBN 9781935623915 (cloth : alk. paper)
Subjects: LCSH: United States—History—Civil War, 1861–1865—Technology. | Technology—United States—History—19th century.
Classification: LCC E468.9 .A88 2016 | DDC 973.7—dc23 LC record available at http://lccn.loc.gov/2015048163

ISBN (print):   978-1-935623-91-5
ISBN (e-book):   978-1-935623-92-2

Printed in Canada

⊗ The paper used in this publication meets the minimum requirements of the American National Standard for Permanence of Paper for Printed Library Materials Z39.48–1992.

CONTENTS

PREFACE

In November 2012, over the Veterans Day weekend, the National Museum of American History launched its observance of the sesquicentennial of the American Civil War. I organized the opening event, an international symposium on the role of technology in the war. Entitled "Astride Two Worlds: Technology and the American Civil War," it was made possible by the financial and material support of the Smithsonian Institution, the National Museum of American History, and the Division of Armed Forces History. The symposium proved highly successful, with 18 participants presenting papers during the three-day event. This book comprises a selection of eight of those presentations, revised and expanded by their authors into formal articles. As context for these articles, I have provided a bibliographical overview of the subject.

The symposium was intended to address a relatively straightforward question. How did nineteenth-century industrialization and rapidly changing technology affect the course and conduct of the Civil War (1861–1865), which fell almost exactly halfway between the end of the preindustrial Napoleonic Wars (1815) and the onset of the fully industrial Great War (1914)? By the middle of the nineteenth century, industrialization and technological innovation were beginning to alter drastically the character and conditions of warfare as it had been conducted for centuries. Occurring in the midst of these far-reaching changes, the American Civil War straddled two ages; it can justly be labeled both the last great preindustrial war and the first major war of the industrial age. Industrial capacity attained new levels of military significance as transportation improved with the expansion of railroads and steamships. Yet in this as in many other respects, the Civil War was distinctly transitional. Horse-drawn wagons and pack mules still carried the main logistic burden despite the growing significance of railroads as transporters of troops and supplies. Smoothbore artillery still dominated the battlefield despite the advent of rifled artillery. Old-fashioned seamstresses still outnumbered new-fangled sewing-machine operators, to cite only a few examples.

The Civil War in America was the first full-scale war shaped in major ways by the tools and weapons of the Industrial Revolution, but that was not always obvious to those in its midst. Even as battle itself presaged the growing mechanization of war with new weapons that multiplied the range and efficiency at which death could be dealt, commanders still looked to the past for tactics and organization. At the same time, ingenuity and imagination marked the efforts of both sides to devise and apply still newer weapons and techniques to realign old habits with new realities. Some were successful, many were not. The enthusiasm for novelty that marked mid-century America produced more failures than successes and *Astride Two Ages* pays attention to both. For all the passion for the new, much of the soldier's gear and arms had changed little from earlier times. That both Union and Confederate arms production included swords, lances, and pikes underscores the war's transitional nature.

But military technology narrowly defined was only part of the story. Many ostensibly civil technologies that had spread widely during the antebellum years strongly influenced the course of the Civil War. Telegraph and railroad greatly increased the pace at which events moved and combined with other technical changes to vastly extend the scope and deadliness of battle while at the same time compromising its decisiveness. Other factors multiplied the numbers who might be exposed to peril. Mechanizing agriculture fed larger armies; growing industries armed and supplied them; steam power moved them on land and sea. Yet here, as with narrower military technology, novelty hardly ruled unchallenged. Horses still mattered more than steam engines, and disease still killed more soldiers than did weapons.

Astride Two Worlds does not pretend to deal comprehensively with technology and the Civil War, but it does illuminate important aspects of the subject and underlines the often overlooked fact that more innovations failed than succeeded. And while novel technology and new modes of production and transportation significantly shaped the conduct of the Civil War, traditional ways of mustering, arming, provisioning, moving, and sustaining the men who fought were certainly no less important.

Barton C. Hacker
National Museum of American History
8 September 2015

ONE

HOW TECHNOLOGY SHAPED THE CONDUCT OF WAR

Barton C. Hacker

In the second quarter of the nineteenth century military technology entered a period of sustained innovation and transformation that deserves to be termed *revolutionary*. The United States entered the preliminary stages of that nineteenth-century military–technological revolution by the 1830s. The revolution was well under way when the American Civil War began, but it was far from completed. The Civil War was both the last great preindustrial war and the first major war of the industrial age. On the eve of war the United States was not yet a fully industrialized nation, but the process of industrialization was well advanced. This was the context that distinguished the American Civil War from all earlier wars as well as making it a harbinger of wars to come.

Technological innovation has regularly been cited as a hallmark of virtually every aspect of the American Civil War (Bruce 1987; Ross 2000; Moorehead 2004). Lists of technologies and techniques first used, or first used extensively, in the Civil War cover a broad range: mass production in some industries, notably clothing and small arms, and new techniques of food preservation; the supplying and movement of mass armies by railroad and steamboat; photography, telegraphy, various signal devices that used flags and lamps, and aerial observation from fixed balloons; the general use of rifled small arms and the appearance of breech-loading and magazine arms, as well as early machine-gun systems; the normal disappearance of troops behind breastworks and into trenches, along with the use of wire entanglements and trench mortars; land and marine mines, torpedo boats, and submarines; and steam-powered armored warships. That arms production also included swords, lances, and pikes again attests to the transitional nature of the American Civil War. And, railroads notwithstanding, horses and mules remained essential to army logistics, as they had for millennia (Bruce 1989; Hagermann 1992; Ried 2014).

TECHNOLOGICAL REALITIES

Two innovations played particularly important parts in shaping the course of the American Civil War. One transformed battlefield tactics: troops on both sides carried rifled firearms. Only one other among the host of Civil War firsts exerted an equally profound effect, but on strategy and logistics more than tactics: steam power applied in manufacturing and transport.

Springfield Armory's Model 1842 smoothbore musket was the first major product of the recent armory-developed uniformity system (Moller 2011:189–238). In addition to relying on interchangeable parts, it also discarded the venerable flintlock in favor of a percussion caplock. Caplocks eliminated the external ignition that made flintlocks prone to misfire. The cap, attached to the weapon through a sealed port, held an unstable chemical compound; a sharp blow from the hammer detonated the chemical, sending a flame through the port to ignite the gunpowder in the chamber. Percussion ignition greatly enhanced reliability (Blackmore 1983; Moller 2011:180–182). When allied with a practical rifling system, as it soon was, it helped radically transform small arms.

Rifles had existed at least since the fifteenth century. Unquestionably superior to smoothbores in range and accuracy, they suffered from one major drawback for military purposes: bullets had to fit the barrel tightly for rifling to work. Ramming a tight-fitting bullet down a three- or four-foot barrel, as muzzle-loading rifles required, was a hard, slow job, even compared with the none-too-easy task of loading a smoothbore musket. One answer was breech-loading, which, like rifling, had a long history and a serious military drawback: gas escaping from the breech. Only in the later nineteenth century did the introduction of metal cartridges combine with improved manufacturing techniques to attain precision enough to meet military needs for cheap, plentiful, and reliable breechloaders (Peterson 1983; Roads 1983).

One stopgap answer to the rifling problem came to fruition in the 1840s, the brainchild of French army Capt. Claude-Étienne Minié, who built on earlier work by another French army inventor, Gustav Delvigne. The key idea was that a bullet small enough to slip easily down the barrel could then be deformed to grip the rifling. Delvigne proposed tapping the ball with a ramrod to deform the bullet, but Minié suggested something more elegant: the self-expanding bullet. Exploding powder would expand the bullet's hollow base, spreading the lead to fill the barrel and grip the rifling. Although Minié's name became firmly attached to the new bullet, the actual conversion of Minié's largely theoretical solution into a practical device was the handiwork an American, the assistant master armorer at the army's Harpers Ferry Armory, James Henry Burton. Subsequently, Burton played key roles in ordnance production for the Confederacy and still later at England's Enfield arsenal (Tate 2005; Collins 2013).

The U.S. Model 1855 rifle-musket used this new bullet, plus an improved percussion system; the slightly modified Model 1861 became the U.S. Army's standard arm when the Civil War began (Moller 2011:252–259). Its adoption came three decades after Springfield Armory had begun its pioneering development of interchangeable parts manufacturing, the uniformity system (Smith 1977, 1985; Mayr and Post 1981; Tull 2001; Ford 2005). Before the 1830s, Springfield never managed to produce more than 5,000 muskets a year. At such production levels, even with several government and contract armories involved, arming the hundreds of thousands who fought in the American Civil War would have been well-nigh impossible. The Confederate seizure of Harpers Ferry Armory left Springfield as the only government arsenal when war began. Its annual capacity by 1861 stood at roughly 12,000 arms. Substantial purchases abroad marked the opening of the war, and private contracts later augmented domestic arms manufacture, but Springfield remained the Union's major source of rifle-muskets. By 1865, in fact, it had become the world's largest arms factory, with an annual capacity of better than 300,000. In chapter 2, "Yankee Armorers and the Union War Machine," Merritt Roe Smith describes in fascinating detail the remarkable increase in productivity by gun makers at the Springfield Armory.

The Confederacy of necessity relied more on foreign supply but nonetheless managed to build a creditable arms industry almost from nothing, aided by the machinery seized at Harpers Ferry. Much of the credit belonged to Josiah Gorgas, a Pennsylvanian who followed his Alabama-born wife into secession and worked miracles in organizing arms production. As chief of ordnance for the Confederacy, he built an extraordinary system of acquisition, manufacture, and distribution of arms and ammunition (Vandiver 1994; Collins 1999). At their peak in mid-1863, southern armories were producing 28,000 small arms a year, augmented by 7,000 from other sources. Though often short of everything else, Confederate troops never lacked firearms, largely thanks to Gorgas and his talented lieutenants (Tate 2005; Davis 2007; Thomas 2010; Barry and Burt 2012; Collins 2013), an accomplishment matched only by the unfailing supply of gunpowder (Johnston 1990; Curtis 2006; Bragg et al. 2007).

The introduction of rifle-muskets to the battlefield in such numbers transformed the technical framework for the tactics that had defined Napoleonic warfare. Smoothbore artillery with an effective range of 400 yards far outranged smoothbore muskets. On Napoleonic battlefields massed artillery allowed attackers to decimate a defending force, clearing the way for a decisive bayonet assault. When rifled small arms outranged smoothbore guns, as they did in the 1860s, defenders routinely forced artillery back from their most effective range. The use of field glasses and telescopes became ubiquitous as the range of weapons increased (Gaddy 2002; Greivenkamp and Steed 2011:81290S-12–81290S-13; Warner 2014). Against unshaken infantry, the familiar Napoleonic tactics of massed frontal assault could succeed only at terrible cost, and bayonets hardly mattered. Troops still tended to engage at the same distances as they had in the smoothbore era, roughly 100 yards, but without effective artillery preparation defenders could wreak havoc on attackers crossing open ground, especially when the defenders dug in (Falk 1964; Griffith 1989; Katcher 2001a; Bilby 2005; Hess 2008, 2013; Miller et al. 2010; Murray 2014). This was another departure from Napoleonic tactics that soldiers quickly learned. The spade became scarcely less important than the rifle. Before it ended, the Civil War had become an engineer's war. Field fortifications, often neglected early in the war, soon became normal. Troops began disappearing into the ground or behind breastworks at even brief pauses. By the final year, elaborate trench systems converted field operations into siege warfare. In this as so much else, the Civil War in America foreshadowed war to come (Field 2005; Hess 2005).

Rifled field artillery might have restored the offensive advantage, but guns small enough to serve in the field could not then be made strong enough to withstand the high pressure of a full-powder charge. Reduced charges left them lacking the necessary hitting power to make them tactically significant (Katcher 2001b; Konstam 2003; Hazlett et al. 2004; Field 2007). Relative immobility limited bigger guns chiefly to service as fortress, coastal, or naval artillery. Naval ordnance, like other heavy ordnance and field artillery, remained predominantly smoothbore but enjoyed the first notable advances in two centuries. Stimulated by French artillery officer Henri-Joseph Paixhans, especially by his *Nouvelle force maritime* (1822), European navies began to adopt explosive shells of improved design in the 1830s. Eventually, the United States followed suit. Because shells were hollow, they could be larger than solid shot of comparable weight. Paixhans also advocated standardization for naval guns, another innovation soon under way. A Swedish-American naval officer and inventor, John A. B. Dahlgren, learned how to raise the caliber of shell guns from 8 to 11 inches, and Thomas J. Rodman's new founding process made guns with calibers up to 20 inches feasible (Johnson 1989, 1990; Schneller 1996). Explosive shells ended the day of steam warships propelled by paddle wheel. Improved naval guns also promoted the use of armor on ships. The development of heavy artillery, rifled as well as smoothbore, is the subject of chapter 3, "Heavy Artillery Transformed" by Steven A. Walton.

Steam power was essential to the new methods of gun founding and reached much further. Not only the heart of a growing industrial revolution, steam power also transformed both land and water transport. On land, steam-powered transport had far-reaching consequences for American development during the first half of the nineteenth century. Railroads offered a degree of flexibility that waterways could never match. Here the prewar changes were revolutionary. In 1830, the Baltimore and Ohio Railroad put 13 miles of track into service. Competing and expanding lines laid over 9,000 miles of track in the next two decades, mostly in the Northeast. Between 1850 and 1860, another 22,000 miles of track extended the rail net into the South and West, though the Northeast retained the densest network and also benefited from wider use of high-quality rails in standard gauge. When the Civil War began, more than 31,000 miles of track crisscrossed the country east of the Mississippi (Angevine 2004; Schwantes 2008; Marrs 2009; Miner 2010; Thomas 2011). During the same period, American engineers also solved most of the design problems that plagued the early steam locomotive, transforming a relatively slow, short-distance puller of a few hundred pounds into a speedy, long-ranged hauler of tons. They developed a distinctively American machine well suited to continental landscapes and distances (White 1997; Lamb 2003).

Growing with the railroad system was the increasingly ubiquitous telegraph system, without which so large and complex a network would scarcely have been feasible. After a decade developing an electromagnetic telegraph, Samuel F. B. Morse and his partners in 1843 formed the Magnetic Telegraph Company to exploit the invention. A line between Washington and Baltimore was in full operation the following year. Before the end of the decade, Morse's company had been joined by a host of others to create a dense web in the East. Commerce drove the early telegraph system, but by the 1850s the wires had begun spreading west and south along the expanding railroad network. By 1861 more than 50,000 miles of telegraph wire linked every major city east of the Mississippi. Like the railroads, most of the wire network was heavily concentrated in the North, but the Confederacy inherited a working system (Thompson 1972; Sterling et al. 2006:238–242; Hochfelder 2012; Miller 2012).

Overenthusiastic from an economic viewpoint, prewar railroad building acquired strategic significance with the outbreak of war. Understanding fully what railroads meant to strategy, however, took time. Unlike at least some Europeans,

Americans had given little if any thought to rail's military implications. In brief, railroads allowed contending states to marshal their resources swiftly. Large bodies of troops might quickly be shifted to where they were wanted. Perhaps more important, steam-powered logistics promised to maintain much larger armies than would otherwise have been feasible. Speed, both of troop movement and of supply, emerged as the essence of rail's military utility. A defeated army falling back on its railheads might soon be repaired, its casualties replaced, its losses made good. Reinforced and reequipped armies could return to fight again, so long as men and material were available (Wolmar 2010:34–63). Rapidly restored armies robbed battle of its potential for prompt decision. The two Civil War battles most often called decisive, Gettysburg and Vicksburg, were both fought almost as near the war's beginning as its end. They were also greatly prolonged. Vicksburg, of course, was a siege, so its length would not be surprising. But Gettysburg was a single battle lasting three days, a little-noted portent of things to come. Ultimately, one side's exhaustion decided the issue, and that would become true not only of battle but also of war.

Dramatically increased speed of movement came at the cost of freedom of movement; transferring supplies from railheads to front lines still depended on muscle power (Shrader 1987; Hagerman 1988; Armistead 2013). Civil War armies became more and more closely tied to their railroad supply lines, the attack and defense of which came increasingly to dominate strategy and tactics. Building and wrecking rail lines became a major military function, but most railroad operations remained almost exclusively profit-making ventures in private hands. It was not an easy relationship. Railroad managers professing concern for their shareholders and army officers demanding logistic support were often at loggerheads. Although the North succeeded better than the South in imposing a degree of order, the South also had major problems elsewhere (Turner 1992; Pickenpaugh 1998; Clark 2001; Hodges 2009). In this as in so much else, material shortages hampered the Confederate war effort: too little track, not enough rolling stock, and insufficient means to produce or procure more (Lash 1991; Black 1998). Meanwhile the Union was not only able to manage effectively the nation's private network but also to create its own U.S. Military Railroad to support the armies directly (Ward 1973; Meredith and Meredith 1979; Weber 1999). Like the railroads, Union telegraphy expanded during the war, including the establishment of military telegraph corps, while the South's deteriorated (Thompson 1972; Wilhelm 1999; Sterling et al. 2006:238–242; Hochfelder 2012; Miller 2012).

Militarily, railroads mattered chiefly to the final element in logistics, distribution. Equally important, though hardly clear at first to planners on either side, was production. Here, too, steam power had already begun to work massive changes in manufacturing organization before the war. In due course the patent importance of rails led eyes back to the food, supplies, material, and equipment they carried forward to the fighting troops (Lackey n.d.). Eventually, this meant that war enlarged its scope to include not only the armies but also the resources and people that supported and sustained them (Wilson 2002; Breakwell 2010; Thomson 2010). Sherman's extraordinary trail of destruction through Georgia and the Carolinas and the extraordinary logistic organization that supported it (Overby 1992; Sokolosky 2002) presaged a new kind of war, perhaps pointing toward the more total war of the next century but still far short of totality (Paskoff 2008; Brady 2012; Nelson 2012). In this, as in many other ways, the Civil War was distinctly transitional—a surprisingly active intersectional trade, for instance, flourished throughout the war (Johnson 1963; Leigh 2014).

Another foreshadowing of wars to come was the extraordinary expansion of armies. The U.S. Army on the eve of the Civil War had a total strength of 16,000 officers and men, who were organized in paper regiments of infantry companies, cavalry troops, and artillery batteries, all more or less at permanent stations save for occasional short-term training exercises. By 1864, Union and Confederate army numbers alike had increased fiftyfold. The Union command structure had expanded to include real regiments, brigades, divisions, corps, armies, and army groups, as well as a national command authority. Although the Confederate command structure never attained army group or national command levels, it expanded at least fourfold. And far from being permanently stationed, all were regularly on the move, for months and even years at a time. Movement involved enormous flows of information for coordination and support of these forces at every level. In chapter 4, "Information Flows and Field Armies," Seymour E. Goodman discusses the tactical, strategic, and logistic implications of field armies enormously larger than any ever seen in America before and the complex, multifaceted technological transformation in communication involving a hybrid of modified techniques from Napoleonic times along with the most modern, widely available technologies of the emerging industrial era that allowed such large armies to be directed, coordinated, and controlled.

Medical care reflected another kind of transition (Flannery 2014; Hawk 2013; Schmidt and Hasegawa 2009; Dammann and Bollet 2007; Rutkow 2005; Bollet 2002; Freemon 1998). Again, a number of firsts may be cited, notably several precursors of scientific medicine—anesthesia in front-line surgery, artificial limbs, medical photography, an ambulance service, hospital trains, and army pharmaceutical labs (Aldrich 2001; Smith 2001; Hawk 2002; Flannery 2004; Burns 2011;

Hasegawa 2012; Devine 2014). The importance of sanitation and camp hygiene was widely respected, if not always fully implemented. Hygiene became a special focus of the U.S. Sanitary Commission, a civilian forerunner of the Red Cross that advised and assisted the Union armies (Hill 1981/1982; Garrison 1999; Giesberg 2000; Humphreys 2013; Jones 2013). For all that, the Civil War ended just before medicine decisively crossed the line to science: general acceptance of microorganisms as the cause of disease and thus recognition of the value of antisepsis occurred in the 1870s. Improved sanitation and medical care nonetheless sharply altered the ratio of battle-caused to disease-caused deaths. As recently as the Mexican War, that ratio was still 1 to 10, the same that had prevailed in the Revolutionary War; that is, for every soldier killed in action, 10 died of disease. In the Civil War, Union armies achieved a ratio no worse than 1 to 2, and even Confederate armies under much less favorable conditions managed 1 to 3 (Gillett 1987; Dorwart 2009).

These advances seem to have had little if any impact on the care of the hundreds of thousands of horses and mules that served both the Union and Confederate armies throughout the war. Horses were, of course, essential to the cavalry, but the artillery also required them in large numbers to pull the caissons, and so did every other branch of the military. Horses and mules were absolutely essential to the transport of ordnance, quartermaster, and every other kind of materiel from railheads to field armies. Equine casualties far exceeded the human cost of the war, at least partly because of the poor care they often received (Armistead 2013). Just how true this was for the Union cavalry David J. Gerleman makes abundantly clear in chapter 5, "Veterinary Care in the Union Cavalry" Nothing suggests that the Confederacy treated its animals any better, and probably far worse, given its more limited resources (Ramsdell 1930; Faust 2000)

The pace of change was no less rapid in naval affairs than it was on land (Symonds 2012; Hackemer 2014). Steam power afloat posed several intractable problems in the early nineteenth century. Machinery remained bulky and inefficient. Installed above the water line to push paddle wheels, machinery also remained far too exposed for a warship, especially when shell guns appeared. Paddle boxes meant fewer guns, thus lesser weight of broadside. Fuel demands that sharply constrained cruising radius only made matters worse. In a single stroke, screw propulsion seemed to resolve most of the problems. One of the first three U.S. Navy vessels mounting shell guns, the 10-gun sloop *Princeton* launched in 1843, was also the world's first warship driven by a stern-mounted screw propeller. Designed by transplanted Swedish engineer John Ericsson, who also oversaw its building, the 400-horsepower engine turned a six-bladed screw that gave *Princeton* a top speed of 13 knots (Tucker 1989).

The extreme vulnerability of wooden ships to shell guns also promoted the next major changes in naval architecture, iron hulls and armor plate. Here again the United States was a pioneer when in 1842 the U.S. Congress authorized construction of an all-iron armor-clad war steamer. Essentially a private venture that the navy endorsed only reluctantly, the so-called Stevens Battery (after its builders) suffered from bureaucratic delays and rapidly changing technology. Ordnance growing in power outstripped efforts to armor the ship against penetration by shot or shell. Never completed, she eventually went for scrap (Small 2008). When the Civil War began, the Union navy had 18 relatively modern steam-powered, screw-propelled, wooden warships. Iron hulls and armor had yet to enter regular service. Although steam proved indispensable to the Union blockade of the Confederacy, iron and armor were not. The burden of blockade fell on wooden ships, which were built at a furious pace. Armorclads played only an auxiliary role, though an important one, dictated in part by the South's reliance on makeshift armorclads to defend its ports.

Transatlantic and coastal steam-powered vessels had become almost commonplace by the 1860s (Gardiner 1993, chap. 4). Potentially, steam made the resources of Europe more readily available to both sides. Practically, it made the Union blockade of the 3,000-mile Confederate coastline feasible. Steamers could patrol waters off southern ports more closely and more regularly than sailing ships ever could have. The price was frequent recoaling. Union forces seized bases on the Confederacy's coast as much to maintain the blockade as to deny their use to the enemy (Dougherty 2010; Hetherington and Kower 2011; Smith 2011). Although far from impenetrable, the Union blockade was effective enough to make Confederate blockade running, which relied heavily on British-built and often more technically advanced ships than the blockaders, extremely lucrative (Foster 1988; Wise 1991; Konstam 2004). Ultimately, however, what success the Union blockade enjoyed may have owed as much or more to such misguided Confederate policies as the infamous cotton embargo (Lebergott 1983; Ekelund and Thornton 1992, 2001).

The Confederacy enjoyed a cadre of Annapolis-trained officers but began the war without ships. It also lacked factories for armor plate and marine engines, hampering naval defense throughout the war (Still 1969). Yet Confederate efforts to rectify the imbalance produced the war's most famous naval action and its best-known military innovation. Despite (or perhaps because of) its indecisive outcome, the 1862 clash off the southern Virginia coast at Hampton Roads between the Confederate *Virginia* (better known as *Merrimack*, the name of the ship upon whose hull she had been constructed) and the Union *Monitor* marked a historic moment. As the first meeting between steam-driven, ironclad warships, it drama-tized as nothing else could the revolutionary developments in naval technology then under way (Konstam 2002a; Holzer and Mulligan 2006; Quarstein 2006; Field 2008).

Of the two, *Monitor* was by far the most novel, another product of the inventive John Ericsson (Thulesius 2007). For the first time in history, a new weapon was designed, developed, and deployed during the same war. Nearly everything about *Monitor* was new and untried, but she was a remarkable suc-cess and became the namesake for an entire class of warships that persisted for decades (Roberts 1999; Konstam 2002c). Ericsson himself made the detailed drawings that guided the workmen as construction proceeded. Upon *Monitor*'s 124-foot hull was riveted a raft-like deck 172 feet long and 41.5 feet at its wid-est; there was 4.5-inch iron armor on her sides and 1 inch on her deck. Most distinctive was the 140-ton turret amidships, a cylinder 9 feet high and 20 feet across encased in 8-inch armor. Revolving on a spindle cogged to an auxiliary steam engine, the turret could swing two 11-inch Dahlgren guns through a full 360 degrees, a field of fire unblocked by masts, rigging, or other sailing-vessel paraphernalia. *Monitor* relied entirely on her engines, also of Ericsson's design, two decades before most warships gave up their auxiliary sails (Mayer-Sommer 1988; Mindell 2012).

Virginia, in contrast, was built on the remains of a standard wooden hull, armor merely draped over a wooden casemate that housed two pivoted 7-inch rifles plus three 9-inch Dahlgren smoothbores, and two 6-inch rifles on each broadside. She also carried a heavy iron wedge fixed to her bow as a ram, largely useless because her rebuilt engines could manage only four knots. Heavily dam-aged in the battle, she required extensive repairs and then had to be destroyed when Union forces seized Norfolk, Virginia, where she was based. The Confed-eracy never ceased its efforts to develop ironclads but could not overcome the same shortage of resources that had hampered *Virginia* (Konstam 2001; Brooke 2002; Quarstein 2012).

Virginia's guns and armor came from Tredegar Iron Works in Richmond, Virginia. By 1860, the foundry was one of the nation's largest, producing locomo-tives, boilers, cables, naval hardware, and cannon. When war came, by means of slave and free labor, it emerged as the industrial heart of the Confederacy. Trede-gar remained the South's premier source of armor and heavy ordnance throughout the Civil War. Its vital importance to the Confederate war effort made Rich-mond's defense as much a military and economic as a political necessity (Crews 1992; Dew 1999; Squyres 1989; Knowles 2001).

Unsolved design problems largely precluded the use of ocean-going iron-clads by either side. Notoriously unseaworthy, they remained restricted to protected coastal waters. Such shortcomings mattered little on western rivers, where armored gunboats played a key role in the Union conquest of the Mississippi. Steamboats had multiplied on American inland waters before the war, having solved some of the worst problems of river navigation, especially in the West. Fitted with iron plate and guns, they became valuable adjuncts to army operations (Smith 2009; Roberts 2002; Konstam 2002b, d).

TECHNOLOGICAL DREAMS

Confederates pursued other technological solutions to breaking the Union blockade of southern coastline, none very successful (Ragan 2003). Among them was the spar-torpedo, first used extensively in the Civil War. Although this device, like other Confederate attempts to counter Union naval superiority, foundered in part on inadequate material resources, it also suffered from operational shortcomings: a dearth of technical support personnel, the absence of adequate doctrine for its best use, and a shortage of men trained to use it most effectively. In chapter 6, "Confederate Spar-Torpedo Boats," Jorit Wintjes describes the technical short-comings of spar-torpedoes but focuses on its operational defects. Despite Confederate failure—only one Union warship ever suffered a successful spar-torpedo attack and the attacker did not survive—this new weapon exerted considerable influence on European naval thought for more than a decade after the war, until the emergence of the self-propelled torpedo.

If armor proved so valuable for gunboats and warships, why not armor individual soldiers? It was a question that a number of soldiers and officers asked themselves, and they readily found several manufacturers happy to oblige. The idea of personal armor proved to be another unrealized technological dream of the Civil War. Whether or not to wear armor was an entirely personal decision, fraught not only with financial and practical considerations but also with moral issues and self-perceptions. In chapter 7, "Armor, Manhood, and the Politics of Mortality," Sarah Jones Weicksel discusses the many reasons that individual body armor in the Civil War remained a technological byway.

Equally futile were a host of Union and Confederate proposals to expand warfare into the atmosphere. The war's one temporarily successful application of aerial techniques was the Army of the Potomac's deployment of observation balloons under the direction of Thaddeus S. C. Lowe. Lowe, a largely self-taught scientist, remained a civilian throughout his service, which lasted from 1861 to

1863. He oversaw the construction of the balloons, which were made of varnished silk, and devised a portable hydrogen-gas generator to inflate them in the field. Aeronauts in his smaller balloons used signal flags to communicate their observations to the ground, but the larger balloons carried telegraph equipment. The system was entirely feasible with mid-nineteenth century technology, but a civilian balloon corps did not sit well with army officers. Despite the utility of observing enemy dispositions from aloft, ballooning did not survive. The corps was disbanded in 1863 (Scott 2014). One of the reasons it failed, although certainly not the most important, was the success of Confederate efforts to mislead observers with false equipment and devious maneuvers. These are discussed by John A. Macaulay in chapter 8, "'Quaker Gun' versus Observation Balloon."

The idea of a heavier-than-air flying machine, although dismissed as fantasy by most people in antebellum America and in Europe, attracted not only quacks but serious scientists and engineers. By 1861, the United States had granted 13 patents for flying machines, and the exciting new magazine, *Scientific American*, had published more than 80 articles on the subject. The war stimulated further interest in both the North and South. In chapter 9, "Dreams of Aerial Navigation," Tom D. Crouch describes in some detail five wartime projects to build powered flying machines, two Union and three Confederate. None succeeded but the dream lived on.

ASTRIDE TWO AGES

The Civil War in America was the first full-scale war shaped in major ways by the tools and weapons of the Industrial Revolution. Telegraph and railroad greatly increased the pace at which events moved and combined with other technical changes to vastly extend the scope and deadliness of battle while at the same time reducing its decisiveness. New weapons multiplied the ranges at which death could be dealt, but other factors multiplied the numbers who might be killed. Agricultural mechanization contributed to growing productivity, greatly promoted by prewar railroad expansion and then by the war itself (Olmstead 1976; Craig and Weiss 1993; Atack and Margo 2011.) Expanded food production permitted larger armies to be fed; industrial growth permitted larger armies to be armed and supplied; and steam-powered transport permitted larger armies to be deployed and sustained (Craig 1996).

Yet, like the uniformity system of the earlier nineteenth century, the rifle-musket that dominated Civil War battlefields and the railroad–telegraph network that dominated Civil War logistics were in some respects typical products of premodern patterns of technological innovation. Technology, whether civil or military, was chiefly empirical. Vast as the accumulation of technical knowledge had become by the mid-nineteenth century, it was still normally the product of hit-or-miss experiment by craftsmen or tinkerers, laboriously augmented over many years, unevenly developed, and slow to spread. But again like the uniformity system, rifle and railroad development incorporated something more novel, a kind of systematic empiricism that began from the eighteenth century onward to accelerate the pace of change. This trend grew even stronger as the century advanced.

Ingenuity and imagination, rather than science, dominated the efforts of both sides to devise and apply new weapons and techniques during the Civil War. Military-technological innovation still relied chiefly on lone inventors and tended to be a prolonged process. Only one wartime invention—Ericsson's *Monitor*—significantly affected the course of the war. In spawning a new class of warship (Konstam 2002a), it also demonstrated how Northern industry could quickly turn invention into an operating military system, just as it could multiply firearms production. The South's inferior—and declining—industry could not hope to match the North's achievement. But not all areas of Northern industry were equally efficient, nor did all officers welcome innovation. Even such eminently workable and potentially significant weapons as Gatling's guns and Lowe's observation balloons saw very limited wartime service (Hughes 2000; Keller 2008; Scott 2014). Manufacturers may deserve some of the blame, but more belonged to key military officials who believed they had good reasons to oppose the introduction of new-fangled devices at the expense of well-tried military practice with a war in progress (Bruce 1993). Industrial shortcomings and military conservatism would change decisively in the half-century that followed 1865.

Note

This essay draws extensively on two earlier works of mine: Hacker 1998 and Hacker 2007, chap. 2. Both will guide the interested reader to many other works dealing with aspects of Civil War technology. See also Bledsoe 2014.

References

Aldrich, Mark. 2001. Train Wrecks to Typhoid Fever: The Development of Railroad Medicine Organizations, 1850 to World War I. *Bulletin of the History of Medicine*, 75, no. 2(Summer):254–289.

Angevine, Robert G. 2004. *The Railroad and the State: War, Politics, and Technology in Nineteenth-Century America*. Stanford, Calif.: Stanford University Press.

Armistead, Gene C. 2013. *Horses and Mules in the Civil War: A Complete History with a Roster of More Than 700 War Horses*. Jefferson, S.C.: McFarland.

Atack, Jeremy, and Robert A. Margo. 2011. The Impact of Access to Rail Transportation on Agricultural Improvement: The American Midwest as a Test Case, 1850–1860. *Journal of Transport and Land Use*, 4, no. 2(Summer):5–18.

Barry, Craig L., and David C. Burt. 2012. *Suppliers to the Confederacy: British Imported Arms and Accoutrements*. Atglen, Pa.: Schiffer.

Bilby, Joseph G. 2005. *Civil War Firearms: Their Historical Background and Tactical Use*. Cambridge, Mass.: Da Capo Press.

Black, Robert C. 1998. *The Railroads of the Confederacy*. Chapel Hill: University of North Carolina Press. [Originally published by same press, 1952.]

Blackmore, H.L. 1983. "The Percussion System." In *Pollard's History of Firearms*, ed. Claude Blair, pp. 161–187. New York: Macmillan.

Bledsoe, Andrew S. 2014. "Technology and War." In *A Companion to the U.S. Civil War*, ed. Aaron Sheehan-Dean, vol. 1, pp. 540–560. 2 vols. New York: John Wiley and Sons.

Bollet, Alfred J. 2002. *Civil War Medicine: Challenges and Triumphs*. Tucson, Ariz.: Galen Press.

Brady, Lisa M. 2012. *War upon the Land: Military Strategy and the Transformation of Southern Landscapes during the American Civil War*. Athens: University of Georgia Press.

Bragg, C.L., Charles D. Ross, Gordon A. Blaker, Stephanie A.T. Jacobe, and Theodore P. Savas. 2007. *Never for Want of Powder: The Confederate Powder Works in Augusta, Georgia*. Columbia: University of South Carolina Press.

Breakwell, Amy. 2010. A Nation in Extremity: Sewing Machines and the American Civil War. *Textile History* 41, Supplement 1(May):98–107.

Brooke, George M., Jr., ed. 2002. *Ironclads and Big Guns of the Confederacy: The Journal and Letters of John M. Brooke*. Columbia: University of South Carolina Press.

Bruce, Robert V. 1989. *Lincoln and the Tools of War*. Urbana: University of Illinois Press. [Originally published Indianapolis, Ind.: Bobbs-Merrill, 1956.]

———. 1987. *The Launching of Modern American Science, 1846–1876. The Impact of the Civil War*. New York: Alfred A. Knopf.

———. 1993. "The Misfire of Civil War R&D." In *Feeding Mars: Logistics in Western Warfare, from the Middle Ages to the Present*, ed. John A. Lynn, pp. 191–215. Boulder, Colo.: Westview Press.

Burns, Stanley B. 2011. *Shooting Soldiers: Civil War Medical Photography by R.B. Bontecou*. New York: Burns Archive Press.

Clark, John E. 2001. *Railroads in the Civil War: The Impact of Management on Victory and Defeat*. Baton Rouge: Louisiana State University Press.

Collins, Steven G. 1999. System in the South: John W. Mallet, Josiah Gorgas, and Uniform Production at the Confederate Ordnance Department. *Technology and Culture*, 40, no. 3(July):517–544.

———. 2013. Yankee Ingenuity in the South: James Burton and Confederate Ordnance Production. *Vulcan: The International Journal of the Social History of Military Technology*, 1, no. 1(June):1–17.

Craig, Lee A. 1996. "Industry, Agriculture, and the Economy." In *The American Civil War: A Handbook of Literature and Research*, ed. Steven E. Woodworth, pp. 505–514. Westport, Conn.: Greenwood Press.

Craig, Lee A., and Thomas Weiss. 1993. Agricultural Productivity Growth during the Decade of the Civil War. *Journal of Economic History*, 53, no. 3(Sept.):527–548.

Crews, E.R. 1992. The Industrial Bulwark of the Confederacy. *American Heritage of Invention and Technology*, 7, no. 3(Mar.):8–17.

Curtis, William S. 2006. "Unorthodox British Technology at the Confederate Powder Works, Augusta, Georgia, 1862–1865." In *Gunpowder, Explosives and the State: A Technological History*, ed. Brenda J. Buchanan, 239–248. Aldershot and Burlington, Vt.: Ashgate.

Dammann, Gordon, and Alfred J. Bollet. 2007. *Images of Civil War Medicine: A Photographic History*. New York: Demos Medical Publishing.

Davis, Robert Scott. 2007. A Cotton Kingdom Retooled for War: The Macon Arsenal and the Confederate Ordnance Establishment. *Georgia Historical Quarterly*, 91, no. 3(Fall):266–291.

Devine, Shauna. 2014. *Learning from the Wounded: The Civil War and the Rise of American Medical Science*. Chapel Hill: University of North Carolina Press.

Dew, Charles B. 1999. *Ironmaker to the Confederacy Joseph R Anderson and the Tredegar Ironworks*. Richmond, Va.: Library of Virginia. [Originally published New Haven, Conn.: Yale University Press, 1966.]

Dorwart, Bonnie Brice. 2009. *Death Is in the Breeze: Diseases during the American Civil War*. Frederick, Md.: NMCWM (National Museum of Civil War Medicine) Press.

Dougherty, Kevin. 2010. *Strangling the Confederacy: Coastal Operations in the American Civil War*. Havertown, Pa.: Casemate.

Ekelund, Robert B., and Mark Thornton. 1992. The Union Blockade and Demoralization of the South: Relative Prices in the Confederacy. *Social Science Quarterly*, 73, no. 4(Dec.):890–902.

———. 2001. The "Confederate" Blockade of the South. *Quarterly Journal of Austrian Economics*, 4, no. 1(Mar.):23–42.

Falk, Stanley L. 1964. How the Napoleon Came to America. *Civil War History*, 10, no. 2(June):149–154.

Faust, Drew Gilpin. 2000. Equine Relics of the Civil War. *Southern Cultures*, 6, no. 1(Spring):23–49.

Field, Ron. 2005. *American Civil War Fortifications (2): Land and Field Fortifications*. Oxford, UK: Osprey.

———. 2007. *American Civil War Fortifications (3): The Mississippi and River Forts*. Oxford, UK: Osprey.

———. 2008. *Confederate Ironclad vs Union Ironclad: Hampton Roads 1862*. Oxford, UK: Osprey.

Flannery, Michael A. 2004. *Civil War Pharmacy: A History of Drugs, Drug Supply and Provision, and Therapeutics for the Union and Confederacy*. London: Pharmaceutical Products Press.

———. 2014. "Medicine and Health Care." In *A Companion to the U.S. Civil War*, ed. Aaron Sheehan-Dean, vol. 1, pp. 590–607. 2 vols. New York: John Wiley and Sons.

Ford, Robert C. 2005. The Springfield Armory's Role in Developing Interchangeable Parts. *Management Decision,* 43, no. 2:265–277.

Foster, Kevin J. 1988. "Builders vs. Blockaders: The Evolution of the Blockade Running Steamship." In *Global Crossroads and the American Seas*, ed. Clark G. Reynolds, pp. 85–90. Missoula Mont.: Pictorial Histories.

Freemon, Frank R. 1998. *Gangrene and Glory: Medical Care during the American Civil War*. Madison, N.J.: Fairleigh Dickinson University Press; London: Associated Universities Press.

Gaddy, David Winfred. 2002. Notes on U.S. Signal Corps Telescopes, 1860–1865. February 2002. http://home.europa.com/~telscope/signalcp.txt (accessed 5 August 2014).

Gardiner, Robert, ed. 1993. *The Advent of Steam: The Merchant Steamship before 1900*. Annapolis, Md.: Naval Institute Press.

Garrison, Nancy Scripture. 1999. *With Courage and Delicacy: Civil War on the Peninsula. Women and the U.S. Sanitary Commission*. Mason City, Iowa: Savas.

Giesberg, Judith Ann. 2000. *Civil War Sisterhood: The U.S. Sanitary Commission and Women's Politics in Transition*. Boston: Northeastern University Press.

Gillett, Mary C. 1987. *The Army Medical Department, 1818–1865*. Washington, D.C.: Center of Military History, U.S. Army.

Greivenkamp, John E., and David L. Steed. 2011. "The History of Telescopes and Binoculars: An Engineering Perspective." In Novel Optical Systems Design and Optimization XIV, ed. R. John Koshel and G. Groot Gregory, *Proceedings of SPIE* (International Society for Optics and Photonics), 8129:81290S-1–81290S-18.

Griffith, Paddy. 1989. *Battle Tactics of the Civil War*. New Haven, Conn.: Yale University Press.

Hackemer, Kurt Henry. 2014. "Naval Development and Warfare." In *A Companion to the U.S. Civil War*, ed. Aaron Sheehan-Dean, vol. 1, pp. 386–409. 2 vols. New York: John Wiley and Sons.

Hacker, Barton C. 1998. "Science and Technology in the Nineteenth Century." In *A Guide to the Sources of United States Military History: Supplement IV*, ed. Robin Higham and Donald J. Mrozek, pp. 82–117. Hamden, Conn.: Archon Books.

———. 2007. *American Military Technology*. Baltimore, Md.: Johns Hopkins University Press. First published 2006 by Greenwood Press.

Hagerman, Edward. 1988. Field Transportation and Strategic Mobility in the Union Armies. *Civil War History*, 34, no. 2(June):143–171.

———. 1992. *The American Civil War and the Origins of Modern Warfare: Ideas, Organization, and Field Command*. Bloomington: Indiana University Press.

Hasegawa. Guy R. 2012. *Mending Broken Soldiers: The Union and Confederate Programs to Supply Artificial Limbs*. Carbondale: Southern Illinois University Press.

Hawk, Alan. 2002. An Ambulating Hospital; or, How the Hospital Train Transformed Army Medicine. *Civil War History*, 48, no. 3(Sept.):197–219.

———. 2013. Civil War Sesquicentennial: Medicine and the Civil War. *Civil War Book Review* (Summer). http://www.cwbr.com/index.php?q=5511&field=ID&browse=yes&record=full&searching=yes&Submit=Search (accessed 23 November 2014).

Hazlett, James C., Edwin Olmstead, and M. Hume Parks. 2004. *Field Artillery Weapons of the Civil War*. Reprinted Champaign: University of Illinois Press. [Originally published London: Associated University Press, 1988.]

Hess, Earl J. 2005. *Field Armies and Fortifications in the Civil War: The Eastern Campaigns, 1861–1864*. Chapel Hill: University of North Carolina Press.

———. 2008. *The Rifle Musket in Civil War Combat: Reality and Myth*. Lawrence: University Press of Kansas.

———. 2013. Civil War Sesquicentennial: Strategy, Tactics, and Fighting the Civil War. *Civil War Book Review* (Spring). http://www.cwbr.com/index.php?q=5436&field=ID&browse=yes&record=full&searching=yes&Submit=Search (accessed 24 June 2014).

Hetherington, Bruce W., and Peter J. Kower. 2011. Technological Diffusion and the Union Blockade. *Explorations in Economic History*, 48, no. 2(Apr.):310–324.

Hill, Patricia. 1981/1982. Eastman Johnson's "The Field Hospital": The U.S. Sanitary Commission and Women in the Civil War." *Minneapolis Institute of Arts Bulletin*, 65:66–81.

Hochfelder, David. 2012. "'Here the Telegraph Came Forceably into Play': The Telegraph during the Civil War." In *The Telegraph in America, 1832–1920*. pp. 6–31. Baltimore, Md.: Johns Hopkins University Press.

Hodges, Robert R., Jr. 2009. *American Civil War Railroad Tactics*. Oxford, UK: Osprey.

Holzer, Harold, and Tim Mulligan, eds. 2006. *The Battle of Hampton Roads: New Perspectives on the USS Monitor and CSS Virginia*. New York: Fordham University Press.

Hughes, James B. 2000. *The Gatling Gun Notebook: A Collection of Data and Illustrations*. Providence, R.I.: Andrew Mowbray.

Humphreys, Margaret. 2013. *Marrow of Tragedy: The Health Crisis of the American Civil War*. Baltimore, Md.: Johns Hopkins University Press.

Johnson, Ludwell H. 1963. Contraband Trade during the Last Year of the Civil War. *Mississippi Valley Historical Review*, 49, no. 4(Mar.):635–652.

Johnson, W. 1989. Admiral John A.B. Dahlgren (1809–1870): His Life, Times and Technical Work in U.S. Naval Ordnance. *International Journal of Impact Engineering*, 8, no. 4:355–387.

———. 1990. T. J. Rodman: Mid-19th Century Gun Barrel Research and Design for the U.S. Army. *International Journal of Impact Engineering*, 9, no. 1:127–159.

Johnston, James J. 1990. Bullets for Johnny Reb: Confederate Nitre and Mining Bureau in Arkansas. *Arkansas Historical Quarterly*, 49, no. 2(Summer):124–167.

Jones, Marian Moser. 2013. *The American Red Cross: From Clara Barton to the New Deal*. Baltimore, Md.: Johns Hopkins University Press.

Katcher, Philip K. 2001a. *American Civil War Artillery 1861–65 (1): Field Artillery*. Oxford, UK: Osprey.

———. 2001b. *American Civil War Artillery, 1861–1865 (2): Heavy Artillery*. Oxford, UK: Osprey,

Keller, Julia. 2008. *Mr. Gatling's Terrible Marvel: The Gun That Changed Everything and the Misunderstood Genius Who Invented It*. New York: Viking.

Knowles, Anne Kelly. 2001. Labor, Race, and Technology in the Confederate Iron Industry. *Technology and Culture*, 42, no. 1(Jan.):1–26.

Konstam, Angus. 2001. *Confederate Ironclad 1861–65*. Oxford, UK: Osprey.

———. 2002a. *Hampton Roads 1862: First Clash of the Ironclads*. Oxford, UK: Osprey.

———. 2002b. *Mississippi River Gunboats of the American Civil War 1861–65*. Oxford, UK: Osprey.

———. 2002c. *Union Monitor 1861–65*. Oxford, UK: Osprey.

———. 2002d. *Union River Ironclad 1861–65*. Oxford, UK: Osprey.

———. 2003. *American Civil War Fortifications (1): Coastal Brick and Stone Forts*. Oxford, UK: Osprey.

———. 2004. *Confederate Blockade Runner 1861–65*. Oxford, UK: Osprey.

Lackey, Rodney C. (n.d.) Notes on Civil War Logistics: Facts and Stories. U.S. Army Transportation Corps website. http://www.transportation.army.mil/History/PDF/Peninsula%20Campaign/Rodney%20Lackey%20Article_1.pdf (accessed 15 September 2014).

Lamb, J. Parker. 2003. *Perfecting the American Steam Locomotive*. Bloomington: Indiana University Press.

Lash, Jeffrey N. 1991. *Destroyer of the Iron Horse: Joseph E. Johnston and Confederate Rail Transport, 1861–1865*. Kent, Ohio: Kent State University Press.

Lebergott, Stanley. 1983. Why the South Lost: Commercial Purpose in the Confederacy, 1861–1865. *Journal of American History*, 70, no. 1 (June):58–74.

Leigh, Philip. 2014. *Trading with the Enemy: The Covert Economy during the American Civil War*. Yardley, Pa.: Westholme.

Marrs, Aaron W. 2009. *Railroads in the Old South: Pursuing Progress in a Slave Society*. Baltimore, Md.: Johns Hopkins University Press.

Mayer-Sommer, Alan P. 1988. An Historical Case Study of Planning and Control under Uncertainty: The Weapons Acquisition Process for the U.S. Ironclad *Monitor*. *Journal of Accounting and Public Policy*, 7, no. 3 (Autumn):201–249.

Mayr, Otto, and Robert C. Post, eds. 1981. *Yankee Enterprise: The Rise of the American System of Manufacturing*. Washington, D.C.: Smithsonian Institution Press.

Meredith, Roy, and Arthur Meredith. 1979. *Mr. Lincoln's Military Railroads: A Pictorial History of the U.S. Civil War Railroads*. New York: W.W. Norton.

Miller, Frederic P., Agnes F Vandome, and John McBrewster. 2010. *Field Artillery in the American Civil War*. Saarbrücken, Germany: VDM Publishing House.

Miller, John H., II. 2012. "Communication and Innovation in the American Civil War: Comparison of Union and Confederate Implementation of Telegraph Technology." Paper presented at the Smithsonian Institution Civil War Sesquicentennial Symposium, "Astride Two Ages: Technology and the Civil War," Washington, D.C., 9–11 November 2012.

Mindell, David A. 2012. *Iron Coffin: War, Technology, and Experience aboard the USS Monitor*. 2nd ed. Baltimore, Md.: Johns Hopkins University Press.

Miner, H. Craig. 2010. *A Most Magnificent Machine: America Adopts the Railroad, 1825–1862*. Lawrence: University Press of Kansas.

Moller, John D. 2011. *American Military Shoulder Arms*. Vol. 3, *Flintlock Alterations and Muzzleloading Percussion Shoulder Arms, 1840–1865*. Albuquerque: University of New Mexico Press.

Moorehead, Richard D. 2004 Technology and the American Civil War. *Military Review*, 84, no. 3 (May/Jun.):6–63.

Murray, Jennifer M. 2014. "Civil War Tactics." In *A Companion to the U.S. Civil War*, ed. Aaron Sheehan-Dean, vol. 1, pp. 211–230. 2 vols. New York: John Wiley and Sons.

Nelson, Megan Kate. 2012. *Ruin Nation: Destruction and the American Civil War*. Athens: University of Georgia Press.

Olmstead, Alan M. 1976. The Civil War as a Catalyst of Technological Change in Agriculture. *Business and Economic History*, 5:36–50.

Overby, Nick. 1992. "Supplying Hell: The Campaign for Atlanta." *Quartermaster Professional Bulletin*, (Winter):4–7.

Paixhans, Henri-Joseph. 1822. *Nouvelle force maritime: et application de cette force à quelques parties du service de l'armée de terre, ou, essai sur l'état actuel des moyens de la force maritime*. Paris: Bachelier Libraire.

Paskoff, Paul F. 2008. Measures of War: A Quantitative Examination of the Civil War's Destructiveness in the Confederacy. *Civil War History*, 54, no. 1(Mar.):35–62.

Pauly, Roger. 2004. "Rifling: A New Twist on an Old Idea." In *Firearms: The Life Story of a Technology*, pp. 59–78. Westport, Conn.: Greenwood Press.

Peterson, Harold L. 1983. "Breech-loading and Repeating Firearms Other than Revolvers 1810–1870." In *Pollard's History of Firearms*, ed. Claude Blair, 239–258. New York: Macmillan.

Pickenpaugh, Roger. 1998. *Rescue by Rail: Troop Transfer and the Civil War in the West, 1863*. Lincoln: University of Nebraska Press.

Quarstein, John V. 2006. *A History of Ironclads: The Power of Iron over Wood*. Charleston, S.C.: History Press.

———. 2012. *The CSS Virginia: Sink Before Surrender*. Charleston, S.C.: History Press.

Ragan, Mark H. 2003. *Submarine Warfare in the Civil War*. Cambridge, Mass.: Da Capo Press.

Ramsdell, Charles W. 1930. General Robert E. Lee's Horse Supply, 1862–1865. *American Historical Review*, 35, no. 4(July):758–777.

Ried, Brian Holden. 2014. "Logistics." In *A Companion to the U.S. Civil War*, ed. Aaron Sheehan-Dean, vol. 1, pp. 74–94. 2 vols. New York: John Wiley and Sons.

Roads, C.H. 1983. "Firearms Other Than Revolvers and Automatic Pistols 1870–1918." In *Pollard's History of Firearms*, ed. Claude Blair, 239–258. New York: Macmillan.

Roberts, William H. 1999. "The Name of Ericsson": Political Engineering in the Union Ironclad Program, 1861–1863. *Journal of Military History*, 63, no. 4 (Oct.):823–843.

———. 2002. *Civil War Ironclads: The U.S. Navy and Industrial Mobilization*. Baltimore, Md.: Johns Hopkins University Press.

Ross, Charles D. 2000. *Trial by Fire: Science, Technology and the Civil War*. Shippensburg, Pa.: White Mane Books.

Rutkow, Ira M. 2005. *Bleeding Blue and Gray: Civil War Surgery and the Evolution of American Medicine*. New York: Random House.

Schmidt, James M., and Guy R. Hasegawa, eds. 2009. *Years of Change and Suffering: Modern Perspectives on Civil War Medicine*. Edinburgh, UK: University of Edinburgh Press.

Schneller, Robert John. 1996. *A Quest for Glory: A Biography of Rear Admiral John A. Dahlgren*. Annapolis, Md.: Naval Institute Press.

Schwantes, Benjamin Sidney Michael. 2008. "Northern Military Railroad and Telegraph Management during the American Civil War 1861–1865". In Fallible Guardian: The Social Construction of Railroad Telegraphy in 19th-Century America, pp. 111–152. Ph.D. diss., University of Delaware, Newark.

Scott, Joseph C. 2014. 'The Infernal Balloon': Union Aeronautics during the American Civil War. *Army History*, 93(Fall):6–27.

Shrader, Charles R. 1987. "Field Logistics in the Civil War." In *The U.S. Army War College Guide to the Battle of Antietam: The Maryland Campaign of 1862*, ed. Jay Luvaas and Harold W. Nelson, pp. 255–284. Carlisle, Pa.: South Mountain Press.

Small, Stephen C. 2008. The Ship That Couldn't Be Built. *Naval History*, 22, no. 5(Oct.):58–63.

Smith, Andrew F. 2011. *Starving the South: How the North Won the Civil War*. New York: St. Martin's Press.

Smith, George Winston. 2001. *Medicines for the Union Army: The United States Army Laboratories during the Civil War*. Binghamton, N.Y.: Haworth Press. [Originally published Madison, Wisc.: American Institute of the History of Pharmacy, 1962.]

Smith, Merritt Roe. 1985. "Army Ordnance and the 'American System' of Manufacturing, 1815–1861." In *Military Enterprise and Technological Change: Perspectives on the American Experience*, ed. Merritt Roe Smith, pp. 117–173. Cambridge, Mass.: MIT Press.

———. 1977. *Harpers Ferry Armory and the New Technology: The Challenge of Change*. Ithaca, N.Y.: Cornell University Press.

Smith, Myron J. 2009. *Tinclads in the Civil War: Union Light-Draught Gunboat Operations on Western Waters, 1862–1865*. Jefferson, N.C.: McFarland.

Sokolosky, Johnny W. 2002. The Role of Union Logistics in the Campaign of 1865. Master's thesis, U.S. Army Command and General Staff College, Ft. Leavenworth, Kansas.

Squyres, Ted T. 1989. *The Tredegar: Logistical Support in the American Civil War*. Maxwell AFB, Ala.: Air War College.

Sterling, Christopher H., Phyllis W. Bernt, and Martin B.H. Weiss. 2006. *Shaping American Telecommunications: A History of Technology, Policy, and Economics*. Mahwah, N.J.: Lawrence Erlbaum Associates.

Still, William N. 1969. *Confederate Shipbuilding*. Athens: University of Georgia Press.

Symonds, Craig L. 2012. *The Civil War at Sea*. New York: Oxford University Press.

Tate, Thomas K. 2005. *From under Iron Eyelids: The Biography of James Henry Burton, Armorer to Three Nations*. Bloomington, Calif.: AuthorHouse.

Thomas, Dean S. 2010. *Round Ball to Rimfire: A Contribution to the History of the Confederate Ordnance Bureau*. Gettysburg, Pa.: Thomas Publications.

Thomas, William G. 2011. *The Iron Way: Railroads, the Civil War, and the Making of Modern America*. New Haven, Conn.: Yale University Press.

Thompson, Robert Luther. 1972. *Wiring a Continent: The History of the Telegraph Industry in the United States, 1832–1866*. Reprinted New York: Arno Press. [Originally published Princeton, N.J.: Princeton University Press, 1947.]

Thomson, Ross. 2010. The Continuity of Innovation: The Civil War Experience. *Enterprise and Society*, 11, no. 1(Mar.):128–165.

Thulesius, Olav. 2007. *The Man Who Made the Monitor: A Biography of John Ericsson, Naval Engineer*. Jefferson, N.C.: McFarland.

Tucker, Spencer C. 1989. U.S. Navy Steam Sloop *Princeton*. *American Neptune*, 49, no. 2:96–113.

Tull, Bruce K. 2001. Springfield Armory as Industrial Policy: Interchangeable Parts and the Precision Corridor. Ph.D. diss., University of Massachusetts, Amherst.

Turner, George Edgar. 1992. *Victory Rode the Rails: The Strategic Place of the Railroads in the Civil War*. Reprinted Lincoln: University of Nebraska Press. [Originally published Indianapolis, Ind.: Bobbs-Merrill, 1953.]

Vandiver, Frank E. 1994. *Ploughshares into Swords; Josiah Gorgas and Confederate Ordnance*. College Station: Texas A&M University Press. [Originally published Austin: University of Texas Press, 1952.]

Ward, James A. 1973. *That Man Haupt: A Biography of Herman Haupt*. Baton Rouge: Louisiana State University Press.

Warner, Deborah Jean. 2014. Seeing with Both Eyes: A Short Account of Opera Glasses, Field Glasses and Binoculars in the United States. *eRittenhouse*. http://www.erittenhouse.org/artitcles/current-issue-vol -25current-issue-vol-25/seeing-with-both-eyes/ (accessed 15 January 2016).

Weber, Thomas. 1999. *The Northern Railroads in the Civil War, 1861–1865*. Reprint Bloomington: Indiana University Press. [Originally published Westport, Conn.: Greenwood, 1952.]

White, John H. 1997. *American Locomotives: An Engineering History, 1830–1880*. Rev. ed. Baltimore, Md.: Johns Hopkins University Press.

Wilhelm, Pierre. 1999. The Telegraph: a Strategic Means of Communication during the American Civil War. *Revista de Historia de América* No. 124 (Jan.–Jun.):81–98.

Wilson, Harold S. 2002. *Confederate Industry: Manufacturers and Quartermasters in the Civil War*. Jackson: University Press of Mississippi.

Wise, Stephen R. 1991. *Lifeline of the Confederacy: Blockade Running during the Civil War*. Columbia: University of South Carolina Press.

Wolmar, Christian. 2010. "Slavery Loses out to the Iron Road." In *Engines of War: How Wars Were Won and Lost on the Railways*, pp. 34–63. New York: Public Affairs.

TECHNOLOGICAL
REALITIES

YANKEE ARMORERS
AND THE
UNION WAR MACHINE

TWO

YANKEE ARMORERS AND THE UNION WAR MACHINE

Merritt Roe Smith

The 9th of April 1865 was doubtless the darkest day of Gen. Robert E. Lee's storied military career. That afternoon Lee met with Gen. Ulysses S. Grant to surrender his battered Army of Northern Virginia to a much larger and better provisioned Union force consisting of the Army of the Potomac, the Army of the James, and the Army of the Shenandoah. Upon returning to his lines near Appomattox Court House, Lee asked his assistant to draft a document to his troops formally announcing the surrender. Lee read and revised the draft and issued it the following day. "General Orders No. 9" (Lee 1865) stated that "after four years of arduous service, marked by unsurpassed courage and fortitude, the Army of Northern Virginia has been compelled to yield to overwhelming numbers and resources." After expressing to his troops "unceasing admiration" for their "constancy and devotion," Lee ended his communication by bidding them "an affectionate farewell." General Orders No. 9 marked the end of the Confederacy's most fabled fighting force and, for all intents and purposes, the Confederacy itself (Freeman 1935, vol. 4:103–155).

Much has been made of Lee's farewell. To be sure, his observation about the Union army's "overwhelming numbers and resources" was telling—especially in the spring of 1865. To rebel commanders like Gen. Jubal Early (1872a, b), the North's superior human and technological strength accounted for the Confederacy's defeat. This view played an important part in the emergence of the postwar myth of the "Lost Cause" in Southern history (Gallagher 1995; Gallagher and Nolan 2000; Gallagher 2001, 255–276; Towns 2012). But Lee's reference to the North's overwhelming numbers and resources also pointed to significant factual issues. The U.S. Census of 1860 had revealed that the North's white population outnumbered the South's by more than three to one. The census showed another equally important fact, one that was not fully appreciated in the South: the vast preponderance of manufacturing in the United States was located north of the Mason–Dixon Line. Over the years, many historians have acknowledged the importance of human and technological resources in winning the war. And in Ken Burns' (1990) popular film history, *The Civil War*, writer Shelby Foote even goes so far as to say that "the North fought that war with one hand behind its back"— an exaggeration, to be sure. In Foote's view, the North's overwhelming advantage in manpower and resources made all the difference. The side with the "strongest battalions," as Richard Current succinctly put it, won the war (Current 1960).

Not surprisingly, both professional and amateur historians continue to debate the validity of this viewpoint. No consensus yet exists on the subject. This chapter has two primary objectives: first, to understand better the technological and industrial basis of the war and how it affected the conflict's outcome; and, second, to gain a perspective on how the war's technological and industrial nature influenced what happened in postwar America and around the world. I focus on three questions: Exactly how and why did the North possess such an overwhelming technological advantage in 1861? To what extent did this technological advantage make a difference in winning the war? And, finally, how did the North's war machine and its associated management system influence what happened in the United States after the war?

None of these questions has a simple answer. If the history of technology teaches anything, it is that technological change is open-ended and more incremental in nature than revolutionary; it is the achievement of many individuals, not just a heroic few. Furthermore, logistics—the production of supplies and movement to where they are required—is a vital topic. In setting the context for my analysis, I intend to show that the logistics of the American Civil War were the product of specific technological and associated organizational innovations that occurred long before the war began. Nothing was foreordained, but once the war began, the past became prologue, setting the course for a level of production, speed, and scale of operations that was unprecedented. In just four years, the war drew on and greatly expanded the technological achievements of the prewar army and private sector manufacturers in novel ways (Smith 1985; Smith 1977; Raber et al. 2008; Wilson 2006). Once the conflict ended, the stage was set for a new era in American history. Thus the Civil War was a critical juncture in the emergence of modern America. It thrust the country into a much steeper trajectory of industrialization, ultimately making the United States (as one contemporary observer put it) "the greatest nation of the earth" (Dunn 1861–1862:1701).

One of the most striking features of the Civil War is the extent to which the North succeeded in outfitting large armies with standardized machine-made equipment. In an age that had only recently witnessed the dawn of the factory system, how was it possible to feed, clothe, and equip a fighting force that, at its peak, numbered around 1 million men and, over the course of four years, enlisted more than 2.6 million? Yet it happened. By 1864, if not earlier, the Union army was the best equipped and provisioned fighting force in the world (Nevins 1971; McPherson 1982; McPherson 1988; Keegan 2009; Guelzo 2012).

When the Civil War erupted in April 1861, neither side was prepared for the vast scale and deadly intensity of conflict that quickly engulfed the fractured republic. Both sides were insufficiently equipped, a fact that became readily apparent at First Manassas, the first major battle of the war. Owing to the diversity of uniforms, both sides suffered the effects of misidentifying the enemy and friendly fire. And both sides quickly discovered the logistical problems that arose from trying to provide ammunition in the field to troops armed with different kinds of weapons that fired different types of ammunition. By the fall of 1862, if not earlier, the Union army had largely overcome these problems, but the Confederate army continued to suffer from erratic supplies and nonuniform equipment throughout the war. Ultimately, the Confederacy would pay a price for those shortcomings, as Robert E. Lee readily acknowledged in 1865. Resources, in short, made a difference. The side with the strongest battalions indeed held a logistical advantage, as many scholars have argued over the years.

The story of Civil War logistics is a huge topic, so I limit this treatment to the Union side of the story and examine the arms-making operations that existed in the North at the outset of the war, and how they changed into something significantly different. Comparable stories can be told about the manufacture of uniforms, footwear, blankets, canteens, wagons, and all sorts of military equipment (Wilson 2006).

THE ORIGINS OF MASS PRODUCTION IN THE ARMS INDUSTRY

Many people associate the birth of mass production with the electrically powered moving assembly lines that the Ford Motor Company introduced in the early twentieth century. But Ford's advanced production system represented the culmination of a process of incremental invention and development that stretched back well over 100 years (Cowan 1997; Maier et al. 2006, chaps. 6, 9, 10, 12). It began with Oliver Evans' automated grist mill in the 1780s (Ferguson 1980), continued with the work of Eli Terry and others in the American clock industry before and after the War of 1812 (Roberts and Taylor 1994), and moved forward to more technically advanced levels of interchangeable manufacturing with Simeon North, John H. Hall, and other innovators in the small arms industry following the War of 1812 (Smith 1977:184–251). But of all of the advances, the contributions of the United States Armory at Springfield, Massachusetts, arguably loomed largest. In 1794, the federal government established national armories at Springfield and Harpers Ferry, Virginia, but after the War of 1812, Springfield assumed the mantel of "Grand National Armory," as superintendent Roswell Lee noted in his letter of 24 December 1816 to Col. Decius Wadsworth (as quoted in Smith 1977:85)—it was the government's most innovative and proficient arms-making facility of the antebellum period. The armory not only helped develop the so-called American system of interchangeable manufacturing, it also emerged as a key clearinghouse for the dissemination of the new technology to other arms makers and to private firms manufacturing technically related products such as sewing machines (Smith 1977:81, 91, 111–112, 285, 288–290, 312–313; Deyrup 1948:118–119). The new methods became well-known among manufacturers as "armory practice" (Hounshell 1984:4–5, 43–46, 49–50, 331–336).

From the end of the War of 1812 through the 1840s, the U.S. Army Ordnance Department oversaw the development of increasingly rigorous gauge-oriented inspection procedures, introduced regulations to govern accounting practices and work procedures on the shop floor, and, in 1839, established a permanent Ordnance Board, comprising senior officers of the Ordnance Department. Among the board's many assignments were the testing of new weapons and materials as well as the publication of special manuals and research reports for the use of army officers and private contractors. Such publications provided detailed specifications (including the machining and gauging sequences used at the Springfield Armory) and inspection and accounting protocols (including the labor and material cost of each component manufactured). The board sponsored scientific experiments on gunpowder, iron, and other materials used in ordnance work. It also dispatched special investigatory teams of officers to Europe to visit armories, arsenals, foundries, and the like to gather information about the latest ordnance developments overseas (Falk 1959). All these prewar activities contributed substantially to the development of a tightly controlled administrative structure that emphasized the Ordnance Department's mantra of "system, order, and uniformity." (Smith 1985:49ff.; Falk 1959:593–594).

During this period, the Ordnance Department extended its "uniformity system" for manufacturing firearms with interchangeable parts to private companies associated with the government's arms contracting system. Well-known private contractors such as Simeon North (Middletown, Connecticut), Lemuel Pomeroy (Pittsfield, Massachusetts), Asa Waters (Millbury, Massachusetts), and Eli Whitney (New Haven, Connecticut) regularly corresponded with and visited Springfield's superintendent. Information gathering was one of the primary purposes of these exchanges, but such cooperation had another dimension. Many of the latest mechanical innovations were invented by private contractors. The Springfield Armory not only introduced these innovations on its shop floor, it also made them available to virtually anyone who visited the premises. This was so because the Ordnance Department insisted that all private arms contractors who wanted to receive government patronage had to share their inventions free of charge with the national armories, and the armories could, in turn, make them available to other private companies. So, for example, Simeon North and John H. Hall, two of the most gifted machine tool inventors during the antebellum period, readily surrendered their proprietary rights in order to continue to receive government contracts (Smith 1977:chap. 8; Smith 1973).

Proprietary knowledge thus became public knowledge as the national armory at Springfield became a central clearinghouse for the dissemination of the new technology to the larger American manufacturing economy. As a result, armory methods spread rapidly into kindred manufacturing areas, becoming what the British referred to as "the American system of manufactures" (Rosenberg 1969; Hounshell 1984, chap. 1). Long before 1861, the Army Ordnance Department had established the practice of cooperation between the public and private sectors that would ultimately play a crucial role during the Civil War.

From this complex tapestry of innovation, cooperation, and diffusion one can discern a genealogy that links the Springfield Armory and New England private arms manufacturers directly to the rise of the machine tool industry in America. Together these institutions and their cadres of remarkably mobile machinists fostered a quasi-educational phenomenon that economic historian Nathan Rosenberg aptly describes as "technological convergence" (Rosenberg 1976:16–17). Armory practice—sometimes wholesale, sometimes in selected segments—soon spread to technically related industries. Thus by the late 1850s, if not earlier, armory methods could be found in factories making sewing machines, pocket watches, padlocks, railway equipment, wagons, and hand tools (Hounshell, 1984, chaps. 2–6). Interestingly, machine tools, fixtures, and production machinery—rather than inspection gauges—constituted the most frequent technological transfers from the arms industry. Since precision manufacturing was expensive, many business owners contented themselves with manufacturing uniform but not necessarily fully interchangeable products. Only the government could afford the luxury of complete interchangeability and thereby foster its development (Smith 1985:78; Howard 1978).

Owing largely to the prewar activities of the U.S. Army Ordnance Department—particularly the efforts of officers like Col. George Bomford as Chief of Ordnance (ca. 1821–1842) and Maj. (later Gen.) James W. Ripley as superintendent at the Springfield Armory (1841–1854)—all the elements of a coordinated system for manufacturing firearms were in place by 1851, well before the onset of the Civil War. The Ordnance Department's emphasis on closely monitored administrative procedures coupled with the new technologies of interchangeable manufacturing set the stage for greatly expanded levels of production during the Civil War.

THE NORTHERN ARMS INDUSTRY, 1861–1865

In 1861, the North held a huge technological advantage over the South in its ability to produce high-quality firearms. Led by the Springfield Armory and the privately owned Colt Patent Firearms Manufacturing Company in nearby Hartford, Connecticut, the northern arms industry included at least 15 private firms, each capable of manufacturing 10,000 (or more) guns annually. The South, by contrast, had no private armory that could equal that level of output. Most producers were relatively small-shop operations turning out several hundred (or fewer) guns a year. The Confederate government armories in Fayetteville, North Carolina, and Richmond, Virginia, were stocked with two partially complete sets of machinery that had been seized in 1861 from the U.S. armory at Harpers Ferry, Virginia. They had the *potential* to manufacture 10,000 to 13,000 guns a year, but owing to erratic supplies of raw materials, an insufficient labor pool of armory workers, and very tight time constraints needed to tool up for large-scale operations, the southern armories never achieved that level of production (Vandiver 1994; Tate 2005; Fuller and Steuart 1944; Davies 2000; Jones 2014; Albaugh and Simmons 1993; Albaugh et al. 1993; Murphy and Madaus 1996; Murphy 2002; Norman 1996; Flayderman 2001:524–242). From the outset of the war, the Confederacy had to rely on privately owned arms, weapons seized from federal arsenals in the South, foreign imports, and battlefield pickups to supplement its scanty production. It simply did not have the resources to match the productivity of its Union adversaries.

During the war years, the Springfield Armory succeeded in making over 805,000 standard rifle-muskets with a labor force that expanded from 545 workers in April 1861 to nearly 3,000 in April 1864 (Deyrup 1948:233). Contrasted

Figure 2.1. **Top:** United States Model 1861 rifle-musket made at the Springfield Armory. A slightly modified version of the U.S. Model 1855, this long-barrelled rifle and its successor, the U.S. Model 1863, became the mainstay small arms of Union infantry troops during the Civil War. (Photo courtesy U.S. National Park Service, Springfield Armory National Historic Site, Springfield, Mass.). **Bottom:** Close-up of the U.S. Model 1861 rifle-musket's lock plate showing Springfield Armory stamping and date of manufacture.

with other small arms used during the Civil War, the relatively cheap, single-shot, muzzle-loading Springfield rifle-musket (Figure 2.1) was the Model T of American firearms: highly standardized with interchangeable parts (Flayderman 2001:464–469; Fuller 1930; Ball 1997; Hartzler et al. 2000). It was much more accurate than the old smoothbore musket that it replaced in 1855. Unlike a smoothbore musket, a rifle-musket had spiral grooves cut along the interior of its barrel. When fired this caused the bullet to spin, giving it a longer, more accurate trajectory. In the hands of a good marksman, the U.S. Model 1855 rifle-musket (the first of its type in the U.S. Army) proved highly accurate at 100 yards, less accurate at 200–300 yards, yet still capable of hitting targets up to and beyond 500 yards, although, as recent research has shown, most battles were fought at much closer quarters (Fuller 1958:58–64; Griffith 1987; Nosworthy 2003; Hess 2008). Yet, improved as it was, the rifle-musket was dated. Compared with the new breech-loading and repeating firearms that came on the market during the late 1850s and early 1860s, the Springfield rifle-musket was slow to load, clumsy to fire, and cumbersome to carry. Still, it had a devoted champion in the U.S. Army's Chief of Ordnance, Gen. James W. Ripley (Tate 2008).

Figure 2.2. General James W. Ripley, former superintendent of the Springfield Armory (1841–1854) and Army Chief of Ordnance (1861–1863). One of the foremost proponents of the "American system" of interchangeable manufacturing prior to the war, Ripley's commitment to the large-scale manufacture of highly standardized single-shot rifle-muskets brought criticism from others committed to the introduction of more advanced breech-loading and repeating rifles during the war. (Photo courtesy U.S. National Park Service, Springfield Armory National Historic Site, Springfield, Mass.)

A graduate of the U.S. Military Academy at West Point, Ripley joined the Army Ordnance Department in 1832 and rose through its ranks to become the first noncivilian superintendent of the Springfield Armory in 1841 (Figure 2.2). A strong proponent of the Ordnance Department's longstanding uniformity policy, he oversaw the reconstruction and retooling of the armory for interchangeable manufacturing during the 1840s, proudly considering this upgrading among his greatest professional achievements. By the time he became Army Chief of Ordnance in 1861, he held unwavering views about the logistical need for uniformity and therefore resisted the introduction of newer, faster-firing, breech-loading, and repeating firearms.

Ripley subsequently came under severe criticism for his seemingly ultraconservative views. Yet recent research has shown that he had good reasons for insisting on the need for a widely distributed, standard-model firearm (Tate 2008). For one thing, he believed that introducing a vast variety of new firearms with different calibers would pose serious logistical problems for the Ordnance Department in distributing ammunition to variously armed troops in the field. He also expressed concern about the nearly unsolvable problem of repairing such an array of damaged arms in the field. None other than Jefferson Davis had recognized these problems during his tenure as Secretary of War. In an address to the Senate on 8 June 1858 (as quoted in Tate 2008:49–50), he observed that newly developed breech-loading guns required special ammunition. . . . As you multiply the varieties of ammunition, you increase the disability of the troops [as] it is impossible for them to get any ammunition which will answer their purpose. If, however, the ammunition be all of one kind, you can go to the first caisson you reach.

In Davis's view, uniformity came first, even if it meant sidetracking the adoption of faster-firing weapons. On this issue, Davis and Ripley were of one mind.

In addition to considering logistical difficulties in the field, Ripley was equally concerned about the time it would take to tool up for the large-scale production of technically complicated and expensive breech-loading weapons. He was well aware that it had taken the Springfield Armory and Britain's Royal Small Arms Factory at Enfield more than four years to retool for the production of new model rifle-muskets during the late 1850s (Deyrup 1948; Tate 2005). "Even Mr. Colt, who has the most complete private armory in the United States or probably elsewhere," Ripley informed Simon Cameron in a letter of 11 June 1861 (quoted in full in Tate 2008:15–17), "states that it will require six months for him to make the first delivery" of standard Springfield rifle-muskets to the army. Ripley concluded that the process of retooling for large-scale production of advanced breech-loading and repeating arms might take many years—years that would place the Union in serious military jeopardy. He had a point, especially given the widespread belief in 1861 that the war would be of short duration.

As much as the Springfield Armory had helped to pioneer the development of interchangeable manufacturing methods prior to 1861 and as much as its personnel was mentally prepared for the daunting task it faced in the spring of 1861, the plant was actually not physically prepared for the demands placed upon it by the Lincoln administration. During the decade of the 1850s, it had averaged 13,643 guns a year—hardly a remarkable record but considered adequate given the size of the regular army and the distribution demands under the Militia Act of 1808. In 1860 it had made only 9,601 rifle-muskets. No wonder tension gripped the War Department when Abraham Lincoln called for 75,000 volunteers in April 1861. That same month, the Virginia militia seized the Harpers Ferry Armory and looted it, a serious blow to the Union that placed a much heavier manufacturing burden on Springfield. After the disastrous Union defeat at First Manassas on 21 July, Lincoln dramatically ratcheted up the call for volunteers to 500,000. These numbers, coupled with the loss of Harpers Ferry, caused considerable concern in Washington and Springfield. Moreover, the demand for increased arms

Figure 2.3. Major Alexander B. Dyer, the administratively talented superintendent of the Springfield Armory from August 1861 to September 1864, when he left the post to head the Ordnance Department. (Photo courtesy U.S. National Park Service, Springfield Armory National Historic Site, Springfield, Mass.)

production continued to grow as calls for Union volunteers mounted further in the late summer and fall of 1861.

To its lasting credit, the Springfield Armory responded unreservedly. Under the capable leadership of Col. Alexander B. Dyer, a West Point graduate and Ripley protégé who was appointed superintendent in 1861, the national armory managed to produce 13,803 rifle-muskets that year (Figure 2.3). Some officials considered this a disappointing figure, but the armory picked up the pace substantially in 1862 after erecting more buildings, converting others to manufacturing purposes, installing more machinery, hiring more workers, introducing night shifts, and subcontracting for parts with local contractors. Production shot up to 102,410 rifle-muskets. By 1863 the armory began to hit its stride with 217,784 units, and in 1864 it reached 276,200, an unheard of number in the world of American arms making. When the war ended in the spring of 1865, the Springfield Armory was on track to produce at least 300,000 rifle-muskets by year's end, probably more. But once hostilities ceased, the War Department rescinded its previous orders, and the armory's production slowed down to a total of 195,341 rifle-muskets for the year. All told, the Springfield Armory ended up manufacturing a grand total of 805,538 rifle-muskets during the war years, an extraordinary production achievement that exceeded the combined output of all other Civil War rifle-musket contractors (Deyrup 1948:233).

Springfield Armory		Ford Motor Company	
Year	Quantity produced	Year	Quantity produced
1859	13,002	1909	13,840
1860	9,601	1910	20,727
1861	13,803	1911	53,488
1862	102,410	1912	82,388
1863	217,784	1913	189,088
1864	276,200	1914	230,788
1865	195,341	1915	394,788
1866	2,405	1916	585,388

Table 2.1. Comparative production statistics by year for the Springfield Armory (1859–1866) and Ford Motor Company (1908–1916). Sources: Springfield (Deyrup 1948: 233); Ford (Hounshell 1984:224).

Was this "mass production"? If one defines the term as the large-scale manufacture of a complex device consisting of interchangeable parts, then, yes, this was an early example of mass production. To be sure, Henry Ford's Model T automobile consisted of many more components than did the Springfield rifle-musket, but few of those automobile parts, if any, were more irregularly shaped and difficult to machine than the interior working components of the rifle-musket's lock or firing mechanism. As Table 2.1 indicates, Springfield's output between 1862 and 1865 compares favorably with Ford's between 1913 and 1916, which were the first three years of the moving assembly line. In 1916, Ford's numbers soared to over 585,000 units and kept climbing. Would Springfield's production numbers have continued to grow had the war continued? Undoubtedly, but no one will ever know what the national armory might have been capable of achieving or when it would have reached its production ceiling. In any case, Springfield's war production record was impressive, especially given the absence of electrical power and the moving assembly-line methods Ford employed 50 years later (Hounshell 1985, chap. 6; Nye 2013, chap. 2).

The Springfield Armory was not the only manufacturing facility to achieve enormous production levels during the Civil War. Samuel Colt's factory in Hartford produced imposing numbers of firearms during the conflict (Edwards 1962, chap. 25; Hosley 1996; Grant 1995). Equally impressive, Henry Burden's Troy Iron and Nail Works near Albany, New York, deployed a line of its owner's patented metalworking machinery to produce 1 million horseshoes a week, reaching an amazing annual output of 51 million.[1] Made from a single piece of wrought iron with no moving parts, horseshoes were admittedly far less complex than firearms. Nevertheless, Burden's wartime production feats attracted considerable attention because they revealed what could be accomplished on a large scale by using self-acting machinery. His achievement had an important catalytic effect on other metalworking firms, helping to accelerate the momentum toward Ford's feat decades later. Although no Civil War manufacturer came close to matching Burden's production capacity, other army contractors rivaled Springfield's numbers, particularly in armaments, clothing, camp equipage, and leather goods (Wilson 2006). One can legitimately conclude that the Civil War was not only an industry-based conflict but also a war that marked the advent of modern mass production.

THE PRIVATE SECTOR

In addition to the government-owned Springfield Armory, a number of private contractors contributed significantly to arming the Union. The largest private producer of small arms was by far the Colt Company. During the war, the Colt plant produced 130,213 revolvers, 115,406 rifles and rifle-muskets, and 3,724 repeating rifles for the U.S. government. This was a far cry from Springfield's remarkable record, but Colt's varied product line included several types of technically complex firearms, such as repeating rifles and handguns with revolving cylinders containing multiple bullet chambers. The Colt Company also signed numerous contracts with state governments while continuing to sell its famous revolvers to private clients. Altogether the value of Colt's contracts with the U.S. government during the Civil War amounted to nearly $4.7 million (Johnson 1868:730–736; Mowbray 2000730–736). The only other arms maker who equaled (and slightly surpassed) Colt's contract level was Robert Parrott of the Cold Spring Foundry near West Point, New York. His contracts for rifled artillery and artillery shells came to an impressive $4.73 million (Mowbray 2000:864–915). Other major manufacturers of artillery during the Civil War were the Ames Manufacturing Company of Chicopee, Massachusetts, Cyrus Alger of Boston, and Charles Knapp of Pittsburgh, Pennsylvania.

More than 30 private firms made small arms for the Union during the war (Mowbray 2000:698–996). After Colt, the largest of the government's small-arms contractors were E. Remington and Sons of Ilion, New York, with over $2.8 million in contracts for hand guns and rifles (Mowbray 2000:922–928; Hatch 1956; Marcot 1998), and the Sharps Rifle Company of Hartford, well-known for its high-quality breech-loading rifles and rifled carbines. Sharps had army contracts worth more than $2.4 million (Mowbray 2000:945–949; Sellers 1978). Civil War arms makers could be found in every northern state. While concentrated mainly in the New England and Mid-Atlantic states, the industry included Midwestern firms such as the Cosmopolitan Arms Company of Hamilton, Ohio. Although not among the largest contractors in the North, this company produced some 9,300 breech-loading "Cosmopolitan" carbines under contract to the Army Ordnance Department. The 6th Illinois Cavalry carried early issues of this carbine on Col. Benjamin Grierson's famous Mississippi raid, a crucial part of Gen. Ulysses Grant's 1863 Vicksburg Campaign (Flayderman 2001:508–509; Rentschler 2014).

Of all the private contractors to the Union army, the firm of Lamson, Goodnow, and Yale best illustrates the level of mechanical talent and versatility that characterized the wartime Northern arms industry (Figure 2.4). Based in Shelburne Falls, Massachusetts, close to the center of the Connecticut Valley arms-making industry, Lamson, Goodnow, and Yale started out in the 1830s as a manufacturer of scythes and cutlery. Its cutlery is still being made today under the brand name of "Lamson Sharp." The key player in the firm was Ebenezer Lamson, a staunch abolitionist whose home became a safe house on the Underground Railroad for slaves escaping to Canada during the antebellum years (Figure 2.5). Blessed with an eye for money-making projects, in 1858 Lamson and his partners purchased the bankrupt armory-machine-shop property of Robbins and Lawrence in Windsor,

Figure 2.4. The Lamson, Goodnow, and Yale machine shop at Windsor, Vermont, originally built by Robbins and Lawrence. This building currently houses the American Precision Museum, which possesses one of the finest collections of nineteenth-century American machine tools in existence. (Photo courtesy American Precision Museum, Windsor, Vt., and Jon Gilbert Fox.)

Figure 2.5. Ebenezer Lamson, senior partner of Lamson, Goodnow, and Yale, proprietors of the former Robbins and Lawrence factory in Windsor, Vermont. Both firms played important roles in the emergence of the commercial machine tool industry in the United States. (Photo courtesy American Precision Museum, Windsor, Vt.)

Vermont, and soon had it up and running for the manufacture of sewing machines (Brown 2012:12–14; Brown 2011; Roe 1916: chap. 15).

This was a logical decision because Robbins and Lawrence (which closed in 1856) had been well-known for its first-rate firearms and machine tools—product lines that were technically related to the manufacture of sewing machines (Rosenberg 1963). Looking back to 1851, six beautifully machined U.S. Model 1841 rifles displayed by Robbins and Lawrence at the London Crystal Palace Exhibition won recognition for their high quality and interchangeable parts. Along with Samuel Colt's impressive display of revolvers, they also attracted the attention of the British government, which sent a special commission across the Atlantic in 1854 to investigate "the Machinery of the United States of America." The commissioners visited dozens of best-practice machine shops, factories, and armories from Boston to Richmond and as far west as Wheeling (Virginia, at that time) and Pittsburgh (Committee on Machinery 1855; Rosenberg 1969). Their primary focus, however, was on the latest developments in arms making at the Springfield Armory, the Colt works, and the Robbins and Lawrence factory, all in New England's Connecticut River Valley. Their final report reflected

MERRITT ROE SMITH

this focus as did their purchases for the Royal Small Arms Factory at Enfield: the British visitors ended up buying over $100,000 in machinery and verifying gauges. Their goal was to retool the Enfield Armory for the production of a rifle-musket that closely resembled the Springfield Armory's famous interchangeable product. One of the primary sellers of this machinery was Robbins and Lawrence, which secured roughly half of the British orders. The resulting "Pattern 53 Enfield rifle" became standard issue not only to British troops but also to Confederate and Union soldiers (along with the Springfield rife-musket) during the Civil War (Tate 2005; Pam 1998; Smithhurst 2011; Barry 2011).

When the Civil War erupted in the spring of 1861, the abolitionist Lamson and his partners stepped up to support the Union. Selling the rights to their sewing machine business, Lamson went to Washington in search of an arms contract and signed two in July 1861 with Gen. Ripley of the Ordnance Department to produce and deliver 50,000 rifle-muskets (Brown 2012:12–25, 30; Mowbray 2000:819–823; Fuller 1958:172–176). Among other things, the contracts stipulated that the guns had to "interchange in their similar parts with each other and with the Springfield muskets" (Mowbray 1998:161; Brown 2012:13–16). Starting in late October 1862, Lamson, Goodnow, and Yale usually made deliveries twice a month and fulfilled their contract in a timely manner, with the final delivery going out on 10 December 1864. The firm had also committed to make several thousand technically advanced breech-loading carbines for the army: the Palmer bolt-action carbine, which was "the first bolt action metallic cartridge arm accepted for U.S. issuance" by the army, and the Ball repeater (Flayderman 2001:506, 517–518; Mowbray 2000:819–823).

If the history of Lamson, Goodnow, and Yale had been limited to its role as one of the 10 largest arms makers during the Civil Way, its story would have been noteworthy but not absolutely critical. But the tale does not end there. The company also played a pivotal role in outfitting other private contractors, as well as the Springfield Armory, with essential arms-making tools and machinery. Thanks largely to the prewar dispersion of former Robbins and Lawrence machinists and the networks that formed around them throughout New England and the Mid-Atlantic states, manufacturers as far away as Baltimore and Washington, D.C., knew about Windsor, Vermont, as a center of metalworking excellence. Robbins and Lawrence and its successor, Lamson, Goodnow, and Yale, were among the first entrants in what would become the Connecticut Valley machine tool industry. This vitally important manufacturing development ultimately made the region nationally and internationally famous as the "Precision Valley."

When war came, other manufacturers naturally turned to Lamson, Good-now, and Yale to replicate and sell the tools and machinery the company was using. Indeed, Lamson's roster of machine tool clients reads like a "who's who" of Civil War gun making. Names such as Whitney, Sharps, Savage, Starr, Remington, Providence Tool, Richardson and Overman, James H. Merrill, Smith and Wesson, Asa H. Waters, Massachusetts Arms Company, New Haven Arms Company (of Henry rifle fame), Trenton Arms Company, Alfred Jenks and Son, Iver Johnson, John Stevens and Company, and even the Springfield Armory appear in Lamson's order book between 1861 and 1863 (Lamson Order Book 1859–1863). In 1861, for example, Springfield purchased 10 drop hammers for its recently erected Water Shops (Figure 2.6). In 1864, the number of drop hammers used for closed-die forging totaled 43 and played an important role in reducing subsequent machining costs at the armory (Raber et al. 2008:228). By the end of the war, the growing machine tool industry would include such well-known firms as Pratt and Whitney of Hartford, Brown and Sharpe of Providence, Rhode Island, and, later, Jones and Lamson of Springfield, Vermont, a successor to Lamson, Goodnow, and Yale (Roe 1916; Broehl 1959; Meyer 2006; Thomson 2009).

Most noteworthy was Lamson, Goodnow, and Yale's impressive repertoire of machinery. From 1861 to 1865, the company was second to none in the machine-making business. The orders received were substantial. In August 1861, for example, the Starr Arms Company of Binghamton, New York, (one of the top 10 private arms contractors for the Union and the manufacturer of a breech-loading carbine and the Starr revolver) ordered 10 milling machines, a barrel-drilling lathe with four spindles, and castings for a pistol-drilling machine. Other orders followed. Also in August 1861, Richard S. Lawrence (the former co-owner of Robbins and Lawrence and then superintendent of the Sharps Rifle Company in Hartford) ordered a nut-boring machine for gun barrels, a second reaming machine for barrels, a letting-in machine for gun stocks, and a set of castings for a broaching machine. The following month, an even larger order came from the Remington factory in Ilion, New York, for milling machines, milling tools, a barrel-drilling machine, a rifling machine, a barrel-polishing machine, and a full set of tools and fixtures for converting older smoothbore muskets into rifle-muskets (Lamson Order Book 1859–1863). After the Colt Company, Remington was the Union's largest small arms contractor during the Civil War with total contracts amounting to nearly $3 million (Mowbray 2000:922–928).

Figure 2.6. When completed ca. 1860, the Springfield Armory's water-turbine-powered "Upper Water Shops" represented a state-of-the-art facility to forge and machine components for what became popularly known during the war as the "Springfield rifle." (Public domain photo from Benton 1878.)

One of Lamson, Goodnow, and Yale's largest orders (and, to my mind, the most revealing) came not from an established arms maker but from a textile and machinery firm. Prior to the war, the Amoskeag Manufacturing Company of Manchester, New Hampshire, had been one of the largest cotton textile manufacturers in America. Early in 1862, the company made a major decision and ended up signing contracts for 6,000 Lindner carbines and 27,001 rifle-muskets with the U.S. Army Ordnance Department. Making the transition from cotton-related products (as well as steam locomotives and fire engines) to firearms involved a significant technological shift. Amoskeag had no background in the manufacture of firearms. Nonetheless, the company's superintending agent, Ezekiel A. Straw, felt confident that Amoskeag possessed sufficient technical expertise to transform itself. He knew that Lamson, Goodnow, and Yale, located some 60 miles away, could supply him with the necessary machinery. Equally important, the Lamson shops were close enough to assist in the retooling process if any bottlenecks occurred. To insure a smooth transition, he recruited Carlos Clark, a skilled Lamson, Goodnow, and Yale machinist, to oversee Amoskeag's arms-making operations (Brown 2012:21–23). Moreover, when Straw signed his contract with the Ordnance Department, he made sure that his firm would be producing the same "special model" rifle-musket that Lamson, Goodnow, and Yale had contracted to make at its plant in Windsor. Amoskeag's contract stipulated that the guns produced had to be "in all respects identical with the standard rifle-musket made at the United States armory at Springfield, Massachusetts, and are to interchange with it and with each other in all their parts" (Mowbray 1998:399).

Between November 1861 and June 1862, Ezekiel Straw placed three large orders for gun-making machinery with Lamson, Goodnow, and Yale. Ranging in type from plain milling and rifling machines to gun-stocking machines, the orders clearly indicate the extent to which Amoskeag relied on Lamson, Goodnow, and Yale for technical know-how (Lamson Order Book 1859–1863; Mowbray 2000:164; Whisker 2002:71–82). With the machines acquired from the Windsor works in place, Amoskeag proceeded to replicate the rifle-muskets under Carlos Clark's supervision. The process went relatively well. The New Hampshire firm delivered its first batch of 501 rifle-muskets in July 1863 and its last batch of 810 in April 1865, just as the war was ending (Mowbray 1998:718–719). Amoskeag's ability to change product lines in little more than a year speaks volumes about the versatility and skill that characterized industrial New England during the war years. Lamson, Goodnow, and Yale's ability to supply such a wide variety of machinery not just to Amoskeag but to many other contractors speaks to the same versatility and skill. Those same qualities would again manifest themselves during World War I and to an even greater degree during World War II when automakers, typewriter firms, and business machine companies—with the critical assistance of machine tool firms—changed product lines and produced everything from rifles and pistols to tanks and airplanes.

Amoskeag was not the only outlier to enter the gun business during the Civil War. Over half of the 21 firms that contracted with the U.S. Army Ordnance Department for Springfield-type rifle-muskets came from backgrounds other than arms making. The largest of these was Alfred Jenks and Son, well-known manufacturers of cotton and woolen textile machinery; their Bridesburg Machine Works in Philadelphia produced over 98,000 rifle-muskets for the army (Flayderman 2001:466). Other than the Springfield Armory itself, Jenks and Son was the largest producer of standard infantry weapons during the war. Their investment in plant and machinery was considerable. In addition to erecting a new building for their arms-making operations, Jenks and Son (like Amoskeag) had to equip their facility with tools and machinery necessary for the undertaking. Accordingly the son, Barton H. Jenks, visited the Springfield Armory and subsequently corresponded with Springfield's master machinist, Cyrus Buckland, about the national armory's manufacturing routines. The younger Jenks dispatched six machinists to Springfield to examine, take notes, and make drawings of the machinery, fixtures, and gauges used there (Jenks Papers n.d.; Mowbray 2000:127). He also purchased a profile milling machine from Lamson, Goodnow, and Yale (Lamson Order Book 1859–1863).

For his part, Cyrus Buckland wrote Barton Jenks on 23 December 1861 (Jenks Papers n.d.), providing him with detailed handwritten notes and documents about the manufacturing processes and inspection procedures used at Springfield. He also invited Jenks to send him various component parts of the rifle-muskets being made at Bridesburg so he could inspect them (using Springfield Armory gauges) and identify various flaws and technical adjustments that needed to be made in order to "get them perfect." While Buckland's appraisals were critical, they ended on a positive note while also encouraging the Philadelphia firm to do its best to meet the high standards demanded by the Ordnance Department.

The sheer volume of Buckland's correspondence with Barton Jenks is very revealing. He clearly invested a lot of time in consulting the national armory's "Work Books" to provide Jenks with detailed information on machining and heat-treating sequences for each component part of the rifle-musket, labor costs per component, materials used, and inspection procedures to be followed before delivering finished guns to the government for final inspection. Buckland provided Jenks with at least 13 documents, many in his own hand. Most notable are his descriptions of manufacturing procedures and inspection practices. Also noteworthy are Buckland's freehand drawings of a barrel-bedding machine for gun stocks and a rolling mill for gun barrels. What is more, he provided Jenks with information on the record-keeping methods used at Springfield. The length and exacting detail of his letters, especially of 26 August and 23 December 1861 (Jenks Papers n.d.), and accompanying technical commentaries reveal the extent to which the national armory was willing to assist fledgling arms contractors gear up for large-scale production. Such correspondence well illustrates how sustained personal contacts and fluid institutional processes facilitated the transfer of advanced manufacturing methods during the war years.

Since no inventory of the Jenks company's gun-making machines can be found, relatively little can be said about where Barton Jenks procured them. Some were doubtless made at the company's Bridesburg shops. Barton Jenks' correspondence with Cyrus Buckland, however, and an 1861 list of "Stocking Machinery" (Jenks Papers n.d.) indicates that the company's gun-stocking machinery came from four outside sources: the American Machine Works in Springfield; George Crompton in Worcester, Massachusetts; and two Philadelphia firms, William Sellers and Company and Bement and Dougherty. The 22 stocking machines listed in their correspondence were based on patterns originally designed by Buckland, owned by the Springfield Armory, and subsequently made available free of charge to machine builders as far south as Philadelphia. They clearly reveal the geographic extent of Springfield's engineering influence during the war.

A detailed inventory of gun-making machinery does exist for a major contractor who came from a background similar to Jenks and Son: William Mason of Taunton, Massachusetts. Like Jenks and Son, Mason made textile machinery (as well as railway locomotives) prior to the war (Lozier 1986; White 1997). After signing a contract with the Ordnance Department in January 1862 for 50,000 rifle-muskets of the standard Springfield Armory pattern, Mason proceeded to equip his plant with gun-making machinery. On 14 June 1862, he prepared a detailed "Statement of Outlays & Obligations on Account of Muskets"[2] that, in effect, constituted an inventory of his gun-making machinery. The list included everything from barrel-turning lathes, rifling machines, and gun-stocking machines to tilt hammers, shafting, and belting. Of the 92 machines listed, the most numerous were 30 milling machines. The total cost of all the equipment was listed at $71,850.00, a large sum at the time. Although Mason doubtless made some machinery at his own Taunton plant, his inventory indicates that he also purchased a considerable portion of what he needed. Among the most expensive items was a set of 15 gun-stocking machines for $24,200. These machines were based on Springfield Armory patterns and Mason acquired them from Wood, Light, and Company and from George Crompton, both located in Worcester. Other orders to replicate the gauges used at the Springfield Armory went to Philos Tyler's American Machine Works in Springfield, a machine shop that had ready access to the national armory's patterns and drawings and did a lucrative business in duplicating them. Using Springfield Armory patterns, the American Machine Works also produced gun-stocking machinery, milling machines, and other items for the Harpers Ferry Armory as well as various arms contractors prior to (and during) the Civil War. (Mowbray 1998:311, 332–333, 340, 392; Smith 1977:332–333, 339, 392, 409).

Although Mason invested a considerable sum in gauges, fixtures, and machinery, like most private contractors he also subcontracted for certain parts of his rifle-muskets. Indeed, the same inventory of June 1862 shows that he had subcontracted for unfinished gun barrels, lock mechanisms, bayonets, and the like for a total of $574,000. His largest order went to the Washburn Company of Worcester for 75,000 "rough" gun barrels costing $150,000. They were shipped to Mason to be turned, rifled, and finished at his own facility. This procedure was a common practice among Civil War gun makers. Records indicate that Washburn subcontracted for over 300,000 rough gun barrels with various private arms contractors during the war (Mowbray 1998:164, 311, 332, 340, 361, 400, 407, especially 341). Writing to the War Department in April 1862, Mason noted that "Messrs. Tyler [American Machine Works] are making my barrel gauges. These I will see compared with the standard gauges at Springfield [Armory]. I have a sample gun and sample parts" (Mowbray 1998:311). Having an accurate set of verification gauges as well as a model gun supplied by the Springfield Armory was essential because Mason's contract, like all those for rifle-muskets, stipulated that the guns he made had "to be in all respects identical with the standard rifle musket made at the United States armory at Springfield, Massachusetts, and are to interchange with it and with each other in all their parts" (Mowbray 1998:309).

Because his investment in machinery and subcontracted parts was so large, Mason sought to increase his government contract from 50,000 to 100,000 rifle-muskets. The War Department, however, had signed too many contracts for too many arms at the outset of the war and refused Mason's request unless he agreed to make all the component parts of the rifle-muskets at his factory in Taunton. The War Department may have used this requirement as a way to deny doubling Mason's contract. But government officials were legitimately concerned that subcontracted parts might not meet the Ordnance Department's strict standard for high-quality interchangeable firearms. Mason could not produce all the parts himself without making an even larger investment in plant and machinery, which would have put him at even greater financial risk. In the end, he did not completely fulfill his existing contract. Although the War Department reduced the number from 50,000 to 30,000, Mason ended up delivering the last of just 28,584 guns on 29 August 1864—1,416 shy of the agreed-upon number. His total income from the government contract came to $596,316.90, considerably less than the $1 million he initially expected to receive (Mowbray 2000:834–836). Given his substantial set-up costs, it is unlikely that he made much profit, if any, on the venture. But he lessened the financial blow by making rifle components as a subcontractor for other arms makers. His largest orders were for finished gun barrels (Mowbray 1998:415).

The story of William Mason's gun-making venture sheds considerable light on the overall picture of Civil War gun making. Many contractors experienced similar financial problems. Some defaulted on their contracts while others, like Mason, fell short on their deliveries to the government. A case in point concerns Charles B. Hoard, a steam engine manufacturer from Watertown, New York. After receiving a contract for 50,000 standard U.S. Model 1861 rifle-muskets in December 1861, Hoard assured the Secretary of War on 8 February 1862 that "I began remodeling my factory, changing and altering my machinery, building such new kinds as I could manufacture in time to meet my contract," adding that "I have also employed an agent to visit Springfield armory and make all necessary plans, drawings, and specifications, to obtain all required instructions to make all the various parts of the guns" (Mowbray 1998:332–333).

Like William Mason, Hoard ended up producing fewer guns than he originally contracted for. Nonetheless, his reliance on the Springfield Armory for technical know-how speaks forcefully to the national armory's inimitable standing in the American arms industry (Mowbray 1998:311, 339–340, 360–365, 392, 409). It was the institutional center of Civil War arms making. Anyone who received a small arms contract from the Chief of Ordnance could, and most likely did, visit the national armory to examine its manufacturing operations and make arrangements for the replication of needed tools and machinery at local machine shops. They relied on Springfield for model arms, verification gauges, and access to its unrivaled collection of machine drawings and patterns. This public–private cooperation with its ready flow of information sharing and technical assistance made a big difference in preparing contractors for interchangeable manufacturing on a large scale. It proved especially helpful to non-arms makers, allowing them to tool up for production much faster than if left to their own devices. As shown in the case of Lamson, Goodnow, and Yale, private armories also shared technical assistance and information among themselves. Overall, inter-armory cooperation helps to account for the remarkable speed with which the arms industry met the Lincoln administration's unprecedented demands for small arms.

An important adjunct of the Civil War arms industry was subcontracting. While the practice had existed long before the Civil War, its prevalence and importance grew substantially during the war years. Subcontracting proliferated despite the fact that the War Department tended to frown on it. Because component parts had to meet the Ordnance Department's stringent standards for overall quality and interchangeability, officials feared that subcontracting would compromise standards. Even so, nearly all arms manufacturers resorted to the practice, including the Springfield Armory, which subcontracted for parts in order to meet urgent War Department calls for more finished weapons.

Although the national armory permitted private contractors and others to visit its workshops, observe its manufacturing operations, make drawings, and borrow machine patterns for replication elsewhere, it was not allowed to make machinery and tools for sale to private clients. Consequently, private manufacturers interested in adopting Springfield's methods had to farm out orders for tools and machinery to various machine shops located in and around Springfield and as far away as Worcester to the east and Windsor to the north. This wartime dispersal of production set the stage for the emergence of a bona fide machine tool industry in New England, a development that became an essential ingredient of America's industrial expansion after the Civil War.

REFLECTIONS AND CONCLUSIONS

From a technological perspective, the Civil War marked a major turning point in American industrial history. Before the war, the United States was largely an agrarian but rapidly industrializing nation that witnessed the advent of the factory system of manufacturing, the introduction of steam engines and steamboats, the construction of roads, canals, and railroads, the invention of telegraphy, and the dissemination of technological knowledge from innovative leaders in the manufacturing economy to other manufacturers who sought more advanced methods of production. An integral part of this inventive prewar age was the development of interchangeable manufacturing in the firearms industry and, closely associated with it, the emergence of a small but vibrant machine tool industry, especially in New England's Connecticut Valley.

New mechanical technologies energized the American economy during the antebellum years. But as pathbreaking as they were, they had yet to reach maturity by 1861; they were still limited in scale and scope. What the Civil War provided was a new and more powerful momentum toward bigness. Feeding off the creative impulses of the prewar years, the war accelerated the northern manufacturing economy's expansion and initiated a movement toward vastly enlarged industrial operations that took the country in a different direction once the conflict ended. These changes became essential components of what many historians refer to as the "second industrial revolution" in America—the America of Andrew Carnegie, Thomas Edison, John D. Rockefeller, and the "age of big business."

Perhaps next to the development and the expansion of the Union's steam-powered ironclad navy, the most significant technological achievement during the war was the advent of mass production, most notably at the Springfield Armory. In addition to achieving unheard-of production levels, the national armory served as a crucial portal through which numerous arms contractors passed while gearing up for wartime manufacturing. The advent of mass production in the firearms industry had at least as much to do with the existence of a well-developed organizational structure capable of coordinating and controlling work as it did with the nuts-and-bolts deployment of new technologies. Much of the credit for fostering such an effective management system belonged to the West Point-trained officers who staffed the U.S. Army Ordnance Department and oversaw its manufacturing and contracting operations (Smith 1985; Hoskin and Macve 1988:45–49, 51).

After the war, the new technologies generated in the arms industry continued to spread and find applications in the production of typewriters, agricultural implements, bicycles, gramophones, cameras, cash registers, automobiles, and a host of other products associated with the mass production industries of the twentieth century (Fitch 1883, 1884; Battison 1976; Rosenberg 1976:9–31; Hounshell 1984). The long-term consequences of these transformations have yet to be fully charted, though many of them clearly need to be examined in light of the Civil War. Already mentioned is the emergence of the modern machine tool industry in the Connecticut Valley, which spread the new technology and machine-based manufacturing to other parts of the American economy, especially to Midwestern cities like Cincinnati and Cleveland, Ohio, and Chicago (Roe 1916; Hubbard 1923–1924). The war also witnessed the passage of the Morrill Land Grant Act of 1862, a key piece of legislation that resulted in the establishment of dozens of land grant colleges around the country. Their primary purpose was to redirect higher education away from traditional classical studies and toward engineering and science, thus meeting the growing demands of the business community for more technically educated employees (Smith 2010).

Another war-related outcome was the upsurge in railroad construction after the conflict, especially the building of the first federally funded transcontinental railroad and the related emergence of the steel industry. The former drew heavily on Civil War engineering construction practices that emphasized speed, scale, and standardization (Shiman 1991; Army 2014). Meanwhile, an increasingly impersonal bureaucratic approach in the management of large-scale industrial enterprises grew out of the war and contributed to mounting labor tensions. As a result, the country saw its first major labor mobilizations and, in the 1870s, a series of widespread strikes. Less known but equally significant was the rise of a massive international trade in small arms and arms-making equipment as well as the formation of gun clubs, shooting competitions, and the National Rifle Association. Such activities had important social, economic, and political implications. Particularly noteworthy is the connection that existed between the emergence of the post–Civil War arms bazaar and the spiraling arms race that preceded World War I.

These are subjects to be explored elsewhere. The point to underscore here is the role of the state, particularly the U.S. military, in setting a course toward a new stage of American industrial development. As industry became much larger in scale and scope, more highly mechanized, and increasingly reliant on science-based education and industrial research, the United States entered the age of big business and vastly expanded markets at home and abroad. These changes proved so different in degree that American industry became different in kind. The same changes set in motion forces that shaped modern urban–industrial America while also propelling the nation toward much more active engagement in world affairs. At the base of this paradigmatic transformation stood the American Civil War.

Did the technical and logistical advances that helped propel the United States into a new age also determine the North's defeat of the South? All told, they were not the sole reason, but they surely were a major reason. The Union's ready access to the tools and management methods of the industrial revolution—particularly those associated with steam transportation and the self-acting machinery that made mass production possible—proved extremely important in bringing about the Confederacy's defeat. In effect, it meant that Union troops were much better equipped and provisioned than were their Confederate adversaries. Such material advantages go a long way toward explaining why the North won. They accord with Jubal Early's postwar contention (quoted in Gallagher 1995:19) that Robert E. Lee's Army of Northern Virginia "had been gradually worn down by the combined agencies of numbers, steam-power, railroads, mechanism, and all the resources of physical science." To focus on Early's statement primarily as an example of "Lost Cause" myth making (as some Civil War historians have done in recent years) is to sidestep a significant point about the industrial nature of the Civil War and its long-term economic consequences.

Yet even if resource advantages proved essential to the Union's victory, they were not sufficient. Other factors entered into the equation, notably military and political leadership, the staying power of soldier morale and civilian commitment, and, above all, events on the battlefield. All these factors—coupled with the many unforeseen contingencies that altered the circumstances of the conflict—are necessary to explain why the North won and the Confederacy lost the Civil War. No single factor is sufficient (McPherson and Cooper 1998; Smith 2006:481482).

Notes

1. I thank Dr. P. Thomas Carroll, director of the Burden Museum in Troy, New York, for providing this information. On the larger significance of Burden's work, see Uselding 1970; Gates 1981; Greene 2004, 2008:138–140.

2. I thank Professor John W. Lozier of Bethany College for providing me with a copy of this document.

References

Albaugh, William A., III, Hugh Benet Jr., and Edward N. Simmons. 1993. *Confederate Handguns*. Wilmington, N.C.: Broadfoot. [Originally published 1963.]

Albaugh, William A., III, and Edward N. Simmons. 1993. *Confederate Arms*. Wilmington, N.C.: Broadfoot. [Originally published 1957.]

Army, Thomas. 2014. Engineering Victory: The Impact of Antebellum America on the Outcome of the Civil War. Ph.D. diss., University of Massachusetts, Amherst.

Ball, Robert W.D. 1997. *Springfield Armory Shoulder Weapons 1795–1968*. Norfolk, Va.: Antique Trader Books.

Barry, Craig L. 2011. *The Civil War Musket*. Warren, Mich.: Watchdog Quarterly.

Battison, Edwin A. 1976. *Muskets to Mass Production: The Men & the Times That Shaped American Manufacturing*. Windsor, Vt.: American Precision Museum.

Benton, James G. 1878. *The Fabrication of Small Arms for the United States Service*. Ordnance Memoranda no. 22. Washington, D.C.: Government Printing Office.

Broehl, Wayne. 1959. *The Precision Valley: The Machine Tool Companies of Springfield, Vermont*. Englewood Cliffs, N.J.: Prentice-Hall.

Brown, Carrie. 2011. Guns for Billy Yank: The Armory in Windsor Meets the Challenge of Civil War. *Vermont History* 79, no. 2(Summer/Fall):141–161.

———. 2012. *Arming the Union: Gunmakers in Windsor, Vermont*. Windsor, Vt.: American Precision Museum.

Burns, Ken. 1990. "Most Hallowed Ground, 1864." Episode 7 of *The Civil War*. Aired 26 September 1990, Public Broadcasting Service.

Committee on the Machinery (of the United States of America). 1855. Report of the Committee on the Machinery of the United States of America. *British Sessional Papers*, 50:1–87.

Cowan, Ruth Schwartz. 1997. *A Social History of American Technology*. New York: Oxford University Press.

Current, Richard N. 1960. "God and the Strongest Battalions." In *Why the North Won the Civil War*, ed. David Donald, pp. 21–37. New York: Simon and Schuster.

Davies, Paul J. 2000. *C.S. Armory Richmond: A History of the Confederate States Armory, Richmond, Virginia and the Stock Shop at the C.S. Armory, Macon, Georgia*. Carlisle, Pa.: Paul J. Davies.

Deyrup, Felicia J. 1948. *Arms Makers of the Connecticut Valley: A Regional Study of the Economic Development of the Small Arms Industry, 1798–1870*. Northampton, Mass.: Smith College Studies in History, 33.

Dunn, W. McKee. 1861–1862. *Congressional Globe*, 37th Cong., 2nd Sess., 2 Dec. 1861 to 17 July 1862, p. 1701 (as quoted in Richardson 1997:1).

Early, Jubal A. 1872a. "Address of General Jubal A. Early." In *Proceedings of the Third Annual Meeting of the Survivors' Association, of the State of South Carolina; and the Annual Address by Jubal A. Early, Delivered before the Association, November 10, 1871*. Charleston, S.C.: Walker, Evans, and Cogswell.

———. 1872b. *The Campaigns of Robert E. Lee*. An address at Washington and Lee University, 19 January 1872. Baltimore: John Murphy.

Edwards, William B. 1962. *Civil War Guns. The Complete Story of Federal and Confederate Small Arms: Design, Manufacture, Identification, Procurement, Issue, Employment, Effectiveness, and Postwar Disposal*. Harrisburg, Pa.: Stackpole.

Falk, Stanley L. 1959. Soldier-Technologist: Major Alfred Mordecai and the Beginnings of Science in the United States Army. Ph.D. diss., Georgetown University, Washington, D.C.

Ferguson, Eugene S. 1980. *Oliver Evans: Inventive Genius of the American Industrial Revolution*. Wilmington, Del.: Hagley Museum and Library.

Fitch, Charles H. 1883. Report on the Manufactures of Interchangeable Mechanism. In *Tenth Census of the United Sates (1880)*. Vol. 2, *Report on the Manufactures of the United States*. Washington, D.C.: Government Printing Office.

———. 1884. The Rise of a Mechanical Ideal. *Magazine of American History*, 11, no. 6 (June).

Flayderman, Norman. 2001. *Flayderman's Guide to Antique American Fire Arms*. 8th ed. Iola, Wisc.: Krause.

Freeman, Douglas Southall. 1935. *R.E. Lee: A Biography*. 4 vols. New York: Charles Scribner's Sons.

Fuller, Claud E. 1930. *Springfield Muzzle-loading Shoulder Arms: A Description of the Flint Lock Muskets, Musketoons and Carbines and the Muskets, Musketoons, Rifles, Carbines and Special Models from 1795 to 1865 with Ordnance Office Reports, Tables and Correspondence and a Sketch of Springfield Armory*. New York: Francis Bannerman Sons.

———. 1958. *The Rifled Musket*. Harrisburg, Pa.: Stackpole.

Fuller, Claud E., and Richard D. Steuart. 1944. *Firearms of the Confederacy*. Lawrence, Mass.: Quarterman.

Gallagher, Gary W. 1995. *Jubal A. Early, the Lost Cause, and Civil War History: A Persistent Legacy*. Milwaukee, Wisc.: Marquette University Press.

———. 2001. *Lee and His Army in Confederate History*. Chapel Hill: University of North Carolina Press.

Gallagher, Gary W., and Alan T. Nolan, eds. 2000. *The Myth of the Lost Cause and Civil War History*. Bloomington: Indiana University Press.

Gates, Arnold. 1981. For Want of a Shoe: Burden's Machine. *Civil War Times Illustrated*, 20, no. 5(Aug.):18–19.

Grant, Ellsworth S. 1995. *The Colt Armory: A History of Colt's Manufacturing Company, Inc*. Lincoln, R.I.: Mowbray.

Greene, Ann Norton. 2004. "War Horses: Equine Technology in the American Civil War." In *Industrializing Organisms: Introducing Evolutionary History*, ed. Susan R. Schrepfer and Philip Scranton, pp. 143–165. New York: Routledge.

———. 2008. *Horses at Work: Harnessing Power in Industrial America*. Cambridge, Mass.: Harvard University Press.

Griffith, Paddy. 1987. *Battle Tactics of the Civil War*. New Haven, Conn.: Yale University Press.

Guelzo, Allen C. 2012. *Fateful Lightning: A New History of the Civil War and Reconstruction*. New York: Oxford University Press.

Hartzler, Daniel D., Larry W. Yantz, and James B. Whisker, 2000. *The U.S. Model 1861 Springfield Rifle-Musket: As Manufactured by the United States Armory at Springfield, Massachusetts, and Various Private Contractors*. State College, Pa.: Tom Rowe.

Hatch, Alden. 1956. *Remington Arms: An American History*. New York: Rinehart.

Hess, Earl J. 2008. *The Rifle Musket in Civil War Combat: Reality and Myth*. Lawrence: University Press of Kansas.

Hoskin, Keith W., and Richard H. Macve. 1988. The Genesis of Accountability: The West Point Connection. *Accounting, Organizations and Society* 13, no. 1:37–73.

Hosley, William. 1996. *Colt: The Making of an American Legend*. Amherst: University of Massachusetts Press.

Hounshell, David A. 1984. *From the American System to Mass Production, 1800-1932: The Development of Manufacturing Technology in the United States*. Baltimore: Johns Hopkins University Press.

Howard, Robert A. 1978. Interchangeable Parts Re-examined: The Private Sector on the Eve of the Civil War, *Technology and Culture* 19, no. 4(Oct.):633–649.

Hubbard, Guy. 1923–1924. Development of Machine Tools in New England. *American Machinist* 59(1923):1–4, 139–142, 241–244, 311–315, 389–392, 463–467, 541–544, 579–581, 919–922; 60(1924):129–132, 171–173, 205–209, 255–258, 271–274, 437–441, 617–620, 875–878, 951–954; 61(1924):65–69, 195–198, 269–272, 313–316, 453–455.

Jenks Papers. (n.d.) "The Barton H. Jenks Papers." Hagley Museum and Library, Wilmington, Del.

Johnson, Andrew. 1868. Message from the President of the United States in answer to a Resolution of the House of 15th March last, asking for information concerning the ordnance department and its transactions, Jan. 14, 1868. U.S. Cong., H.R., Committee on Ordnance, 40th Cong. 2nd Sess., Exec. Doc. No. 99. Reproduced in Mowbray 2000.

Jones, Gordon L. 2014. *Confederate Odyssey: The George W. Wray Jr. Civil War Collection at the Atlanta History Center*. Athens: University of Georgia Press.

Keegan, John. 2009. *The American Civil War: A Military History*. New York: Alfred A. Knopf.

Lamson Order Book. 1859–1863. Lamson, Goodnow, and Yale Order Book, December 1859–September 1863. American Precision Museum, Windsor, Vt.

Lee, Robert E. 1865. General Orders No. 9, 10 April. Reproduced in Freeman 1935, vol. 4, pp. 154–155.

Lozier, John W. 1986. *Taunton and Mason—Cotton Machinery and Locomotive Manufacture in Taunton, Massachusetts 1811–1861*. New York: Garland.

Maier, Pauline, Merritt Roe Smith, Alexander Keyssar, and Daniel J. Kevles. 2006. *Inventing America: A History of the United States*. 2nd ed. New York: W.W. Norton.

Marcot, Roy. 1998. *Remington, "America's Oldest Gunmaker."* Peoria, Ill.: Primedia.

McPherson, James M. 1982. *Ordeal by Fire: The Civil War and Reconstruction*. 2nd ed. New York: McGraw-Hill.

———. 1988. *Battle Cry of Freedom: The Civil War Era*. New York: Oxford University Press.

McPherson, James M., and William J. Cooper. 1998. *Writing the Civil War: The Quest to Understand*. Columbia: University of South Carolina Press.

Meyer, David R. 2006. *Networked Machinists: High-Technology Industries in Antebellum America*. Baltimore: Johns Hopkins University Press.

Mowbray, Stuart C., ed. 1998. *Civil War Arms Makers and Their Contracts: A Facsimile Reprint of the Report by the Commission on Ordnance and Ordnance Stores, 1862*. Lincoln, R.I.: Andrew Mowbray.

———. 2000. *Civil War Arms Purchases and Deliveries: A Facsimile Reprint of the Master List of Civil War Weapons Purchases and Deliveries Including Small Arms, Cannon, Ordnance and Projectiles*. Lincoln, R.I.: Andrew Mowbray. Reproduction of Johnson 1868.

Murphy, John M. 2002. Confederate Carbines and Musketoons. Santa Ana, Calif.: Graphic.

Murphy, John M., and Howard M. Madaus. 1996. *Confederate Rifles and Muskets: Infantry Small Arms Manufactured in the Southern Confederacy, 1861–1865*. Newport Beach, Calif.: Graphic.

Nevins, Allan. 1971. *The War for the Union*. Vol. 3, *The Organized War, 1863–64*. New York: Charles Scribner's Sons,

Norman, Matthew W. 1996. *Colonel Burton's Spiller & Burr Revolver: An Untimely Venture in Confederate Small-Arms Manufacturing*. Macon, Ga.: Mercer University Press.

Nosworthy, Brent. 2003. *The Bloody Crucible of Courage: Fighting Methods and Combat Experience of the Civil War*. New York: Carroll and Graf.

Nye, David E. 2013. *America's Assembly Line*. Cambridge, Mass.: MIT Press.

Pam, David. 1998. *The Royal Small Arms Factory, Enfield, and Its Workers*. Enfield, UK: David Pam.

Raber, Michael Scott, Patrick M. Malone, Robert B. Gordon, and Carolyn C. Cooper. 2008. *Forge of Innovation: An Industrial History of the Springfield Armory, 1794–1968*. Ed. Richard Colton. Springfield, Mass.: Eastern National.

Rentschler, Thomas B. 2014. "Woefully Deficient in Cavalry Arms": Ohio-made Carbines in the Civil War. *Timeline*, 31, no. 1 (Jan.–Mar.), 44–49.

Richardson, Heather. 1997. *The Greatest Nation of the Earth: Republican Economic Policies during the Civil War*. Cambridge, Mass.: Harvard University Press.

Roberts, Kenneth D., and Snowden Taylor. 1994. *Eli Terry and the Connecticut Shelf Clock*. 2nd ed. Fitzwilliam, N.H.: Ken Roberts.

Roe, Joseph W. 1916. *English and American Tool Builders*. New Haven, Conn.: Yale University Press.

Rosenberg, Nathan. 1963. Technological Change in the Machine Tool Industry, 1840–1910. *Journal of Economic History*, 23, no. 4(Dec.):414–443. Reprinted in Rosenberg 1976:9–31.

———. 1969. *The American System of Manufactures: The Report of the Committee on the Machinery of the United States 1855, and the Special Reports of George Wallis and Joseph Whitworth 1854*. Edinburgh, UK: Edinburgh University Press.

———. 1976. *Perspectives on Technology*. New York: Cambridge University Press.

Sellers, Frank. 1978. *Sharps Firearms*. North Hollywood, Calif.: Beinfeld.

Shiman, Philip. 1991. Engineering Sherman's March: Army Engineers and the Management of Modern War, 1862–65. Ph.D. diss., Duke University, Durham, N.C.

Smith, Merritt Roe. 1973. John H. Hall, Simeon North, and the Milling Machine. *Technology and Culture*, 14, no. 4(Oct.):573–591.

———. 1977. *Harpers Ferry Armory and the New Technology*. Ithaca, N.Y.: Cornell University Press.

———. 1985. "Army Ordnance and the 'American System' of Manufacturing, 1851–1861." In *Military Enterprise and Technological Change*, ed. Merritt Roe Smith, pp. 39–87. Cambridge, Mass.: MIT Press.

———. 2006. "Civil War (1861–1865)." In *Inventing America: A History of the United States*, 2nd ed., ed. Pauline Maier, Merritt Roe Smith, Alexander Keyssar, and Daniel J. Kevles, pp. 451–483. New York: W.W. Norton

———. 2010. "'God Speed the Institute': The Foundational Years." In *Becoming MIT: Moments of Decision*, ed. David Kaiser, pp. 15–36. Cambridge, Mass.: MIT Press.

Smithhurst, Peter. 2011. *The Pattern 1853 Enfield Rifle*. Oxford, UK: Osprey.

Tate, Thomas K. 2005. *From under Iron Eyelids: The Biography of James Henry Burton*. Bloomington, Ind.: AuthorHouse.

———. 2008. *General James Wolfe Ripley, Chief of Ordnance: Answers to His Critics*. North Charleston, S.C.: Booksurge Publishing.

Thomson, Ross. 2009. *Structures of Change in the Mechanical Age: Technological Innovation in the United States, 1790–1865*. Baltimore: Johns Hopkins University Press.

Towns, W. Stuart. 2012. *Enduring Legacy: Rhetoric and Ritual of the Lost Cause*. Tuscaloosa: University of Alabama Press.

Uselding, Paul J. 1970. Henry Burden and the Question of Anglo-American Technological Transfer in the Nineteenth Century. *Journal of Economic History*, 30, no. 2(June):312–337.

Vandiver, Frank E. 1994. *Ploughshares into Swords: Josiah Gorgas and Confederate Ordnance*. College Station: Texas A&M University Press. [Originally published 1952.]

Whisker, James B. 2002. *U.S. and Confederate Arms and Armories during the American Civil War*. Vol. 1, *U.S. Rifles and Rifle-Muskets of the Civil War*. Lewiston, N.Y.: Edwin Mellen Press.

White, John H., Jr. 1997. *American Locomotives: An Engineering History, 1830–1880*. Baltimore: Johns Hopkins University Press.

Wilson, Mark R. 2006. *The Business of Civil War: Military Mobilization and the State, 1861–1865*. Baltimore: Johns Hopkins University Press.

THREE	HEAVY ARTILLERY TRANSFORMED
	Steven A. Walton

*In few other manufactures has it been found necessary to
search so deeply into the materials nature provides in order to
find out the best and strongest, and then to apply it skillfully
[sic], so as fully to develop its strength, as in the manufacture
of guns.*
(Anonymous 1870:69)

F rom the opening shots at Fort Sumter to the annihilating fire from Little
Round Top against Pickett's men and the months of bombardment at Petersburg, artillery played a role not really seen in American experience before the
Civil War. The war against Mexico during 1846–1848 had certainly set the stage,
with tactical innovations of the "flying artillery" and the combined tactics of the
bombardment at Veracruz, but—to paraphrase Tennyson (1870)—the Civil War
fully saw cannon to the left, cannon to the right, and the jaws of hell filled with
storms of shot and shell.[1] In the half century before the war, the U.S. ordnance
arsenal expanded from a set of traditional smoothbore guns defined by their shot
size—6-pounder, 12-pounder, 24-pounder, and 32-pounder being the most common—to a mixture of smoothbore and rifled guns of different types, most known
by their inventor's names: Parrott guns, Dahlgren guns, Rodman guns, and so on.
There were at least another half dozen other well-known inventions in cannon
in the decades preceding the war, including Stockton guns, Blakely rifled guns,
and Armstrong muzzle- and then breechloaders, and the new patterns by Wiard,
James, Treadwell, and Woodbridge. It is generally agreed that by the time the war
was over the age of "modern" built-up guns had arrived, although this revolution

is often improperly reduced to only the developments made by the English inventor William Armstrong at the end of the 1850s (Stoney 1870a; Bastable 1992). Suffice to say, however, that when the United and Confederate States went to war, they went to war with a mélange of old and new and two ordnance corps well versed in the problems of artillery of the day.

The rush to war was not smooth, however, and artillery played a role quite beyond its direct destructive potential. In part this was due to the unparalleled and hurried expansion of the army: the Army of the Potomac, for example, swelled from 9 batteries of 650 men and 39 guns to 92 batteries with 12,500 men and 520 guns within 8 months at the start of the war (Tidball 2011:6). When a British delegation from Canada visited in 1862 and reported back to London, they wrote that "the opening of the civil war in The United States found the opposing parties ill supplied" with field artillery, but because of the size of the armies and the raw nature of the recruits, "a considerable force of artillery was required to support and give confidence to them" (Williams 1862:9). The delegation found that McClellan's Army of the Potomac alone had over a dozen different types of guns in service. The problem was that as the "improvements of the age" had shown the desirability of rifled guns, numerous systems, "of more or less merit," were proposed and tried by the government, despite the strident warnings of field, ordnance, and artillery officers of the imprudence in bringing such a varied set of models into use at one time. Some of this variety that "caused great confusion and vexation" arose because much of the field artillery came with the state volunteer units who "ransacked . . . the arsenals of the North . . . and every piece, of whatever character, was placed in the hands of the troops then taking the field" (Tidball 2011:4). The problem was so acute that some batteries (of only four to six guns) had two or three types of pieces in service at once. In addition, as Henry J. Hunt, Chief of Artillery for the Army of the Potomac, put it (Hunt 1891:400),

> the complication from which the [3-pounder] Napoleon gun [adopted in 1857] had relieved us,—a great variety of ammunition,—was brought back with the rifle-gun, for which different systems of projectile, Parrott's, Shenkl, Hotchkiss, and Ordnance, were supplied, [each of] which gave different ranges with the same charge of powder. These systems would get mixed in the same battery, and affect its efficiency.

The British were less critical about the varieties of siege and garrison artillery, though there, too, they found over a dozen types in use (Williams 1862:15–16).

It is a near truism that the American Civil War ushered in modern, industrialized war with railroads, telegraphs, the Minié ball, floating harbor mines (then known as torpedoes), ironclads, and, especially, rifled heavy artillery (Bruce 1956; Ross 2000; Army 2016). The arrival of near-total war and a glimpse of the trench warfare that would manifest itself half a century later in World War I also defined the conflict from 1861 to 1865 as a new beginning. That new era of warfare, however, was preceded by decades of development in which artillerists and ordnance manufacturers confronted known and perceived limitations in their guns and developed a number of different types of solutions to advance their art.

As early as the 1830s, American civilians, military officers, and ordnance producers began to reconsider the value of artillery, and even without the motivation of the Civil War itself they began investing heavily in its development. Artillery underwent numerous and rapid changes in the decades leading up to the Civil War. The guns that Admiral Perry used on Lake Erie in the War of 1812 were not at all different than the ones Francis Drake had aboard the *Golden Hind* on the Spanish Main, and the guns Maj. Gen. Andrew Jackson unleashed on Gen. Edward Pakenham at the Battle of New Orleans were little different in form, though now mostly made of iron, as those captured at Ticonderoga in 1777 or, for that matter, at the siege of Ostende in 1600. But between 1830 and 1860, numerous new types of guns came into service, and perhaps even more importantly, new manufacturing processes were developed here and abroad that made many of the guns of the Civil War a new breed entirely. Guns became rifled for better accuracy. Shells replaced round shot for accuracy and penetrating power. Guns grew massive: Parrott guns firing 300-pound projectiles were made, other guns grew to 10, 11, or 12-inch bores, and the major foundries were casting 15-inch Rodman guns by 1863. Some trial 20-inch Dahlgren guns were even cast at the end of the war, though these proved too ambitious for use. Quite beyond the matter of guns becoming gargantuan, inventors here and abroad were experimenting with built-up guns, and modern breech-loading artillery was in its early phases. This frothy ferment of the mid-nineteenth century represents a relatively understudied period in the history of artillery but a crucial one for understanding how traditional ordnance could be improved so that it came to be the instrument of change in the Civil War.

INVENTION PRIOR TO THE CIVIL WAR

In America, the War Department was in charge of artillery development for the army, and in the antebellum period the Secretary of War tended to delegate that process to the Chief of Ordnance. The navy similarly had a Bureau of Ordnance (before 1842 part of the Bureau of Naval Commissioners) that oversaw their development and production of artillery. From the Age of Revolution through the Napoleonic era, these boards made contracts with independent foundries to cast guns of standard patterns that fairly closely echoed British and French cannon (Mauncy 1949). These smoothbore guns came in a series of standard sizes from 3-pounder to 32-pounder, each generally a scaled version of the next. But within the decade after the War of 1812 the failures that had always been a feature of cast bronze and iron ordnance under heavy use began to interest many people. Militaries began to try to understand the forms and materials in order to redesign the guns to make them at once stronger and lighter.

Both the army and navy ordnance offices were funneled suggestions and enquires concerning ordnance and, being the efficient bureaucracies they were, were very good at keeping records of the proposals and trials, if they were warranted. A particularly clear set—categorized by type such as general ordnance inventions, breech-loading inventions, and ordnance accouterments (e.g., carriages), as well as small arms, bladed weapons, and so on—are kept in the Records of the [Army] Office of the Chief of Ordnance in the National Archives and show the rate at which inventions were proposed to the army (Figure 3.1). The report register shows decades upon decades of proposals, ideas, and schemes to make new

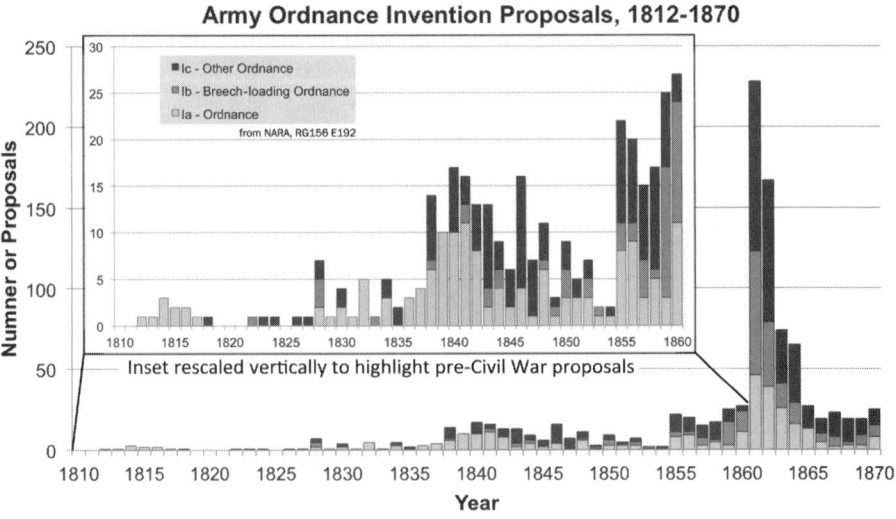

Figure 3.1. United States Army ordnance invention proposals, 1812–1870, classified as general ideas (class Ia), breech-loading apparatus ideas (class Ib), and other ideas related to heavy artillery (class Ic). Compiled by author from NARA (National Archives and Record Administration) RG156/E192.

STEVEN A. WALTON

artillery, indicating that it was not just the Civil War that catalyzed invention. It is clear, though, that as soon as the Civil War began—and in fact even just before the war, for it was clear to everyone that times were tense—many aspiring inventors started sending in ideas for new ways of building or shooting ordnance. Designers proposed all manner of improvements, and the U.S. Army Ordnance Department classified them into general ideas (class Ia), ideas for new breech-loading apparatus (class Ib), and other things related to heavy artillery, such as igniters and carriages (class Ic). The usual quantity of, at most, 20–30 proposals each year before the Civil War was immediately surpassed with well over 200 in the opening year of the war. There is evidence that the Ordnance Department urged the adoption of rifled artillery by the army at the start of the war, and the younger artillery officers all but demanded it "as the latest improvement" (Hunt 1891:400). This initial enthusiasm to help the war effort waned quite steadily with the reality of a continuing war. After the war, proposals settled back in at about two dozen a year.

The massive surge of patriotic invention in crisis of the 1860s, however, drowns out the steady-state background of invention before the war, and looking at only those inventions up to 1860 brings out some interesting patterns in how ordnance was evolving in that period (Figure 3.1, inset). From 1810 through the 1850s there was continuing proposed innovation in traditional muzzle-loading ordnance; the War of 1812 showed the same surge and abatement in invention proposals as the Civil War, though at a much smaller scale. From the late 1830s into the 1840s, though, one can see a consistent push toward new types of basic ordnance, as well as the beginning of interest in breech-loading inventions. In this whole period from 1810 to 1860, a great proportion of these ideas came from within the military itself: lieutenants and captains from the artillery and other branches proposed better ways to do things (during the Civil War the number of civilian proposals was far higher). Here are found, among all manner of ideas that range from plausible to less-than-sound, the revolutionary suggestions of people like Lt. Thomas J. Rodman proposing his water-cooled casting method for cannon, the employer of the newly retired Capt. Robert P. Parrott with the idea of what would become the Parrott Gun, and, on the naval side, Lt. John Dahlgren proposing a soda-bottle-shaped gun whose proportions followed the internal pressures of the gun.

The impetus for development in guns was twofold. First there was the frustrating tendency of guns to burst after long or hard use. Gun failures had become an annoying and sometimes lethal commonplace occurrence as guns and ammunition grew in size and powder manufacture became more reliable (and hence stronger). These failures were a result of what we now understand to be internal

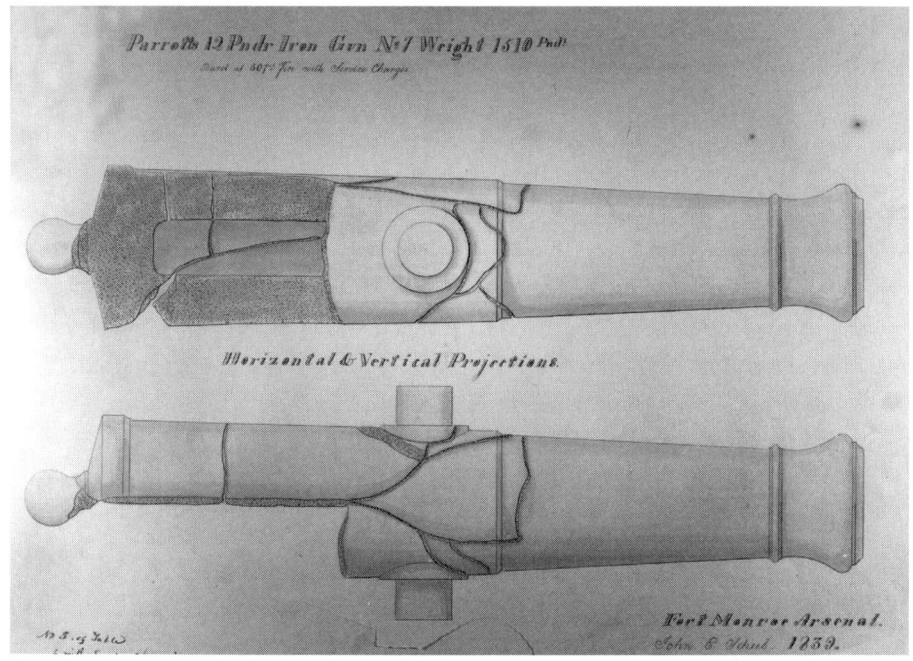

Figure 3.2. Experimental bursting of cast iron gun, 1839. NARA RG156, Portfolio of Drawings, 1814–1870, no. 12. (Document photo by author.)

casting flaws (voids) and fundamental design features (such as sharp corners) that led to stress concentrations within the metal as well as metal fatigue from repeated firing (Figure 3.2). The concept of stress concentrations was known by the time of the Civil War, as is evidenced by Confederate Maj. John Barnwell's, inspection report of James Island in 1863 (Confederate ordnance inspection report [series 1, vol. 28, part 2, serial.47, pp. 378–79], as reported by Swain 2013). In speaking of ruptured guns, Barnwell noted that

> Some manufacturers of ordnance deny the fact that a gun is weakened by rifling . . . [but] it is a fact that the fractures in rifled guns follow the edge of the groove exactly as ice and granite fracture in lines cut upon the surface. It is known that acute re-entering angles upon the surface of guns are the usual lines of rupture, hence the present external form of guns without moldings.

He therefore concluded that "no rifled guns should have acute or sharp-edged grooves," but rather flattened curves, and he singled out the Parrott guns for already having this feature.

STEVEN A. WALTON

Bursting guns were a large liability in human cost, material waste, and operational readiness, so the military was understandably interested in avoiding such losses. A second impetus for gun development, however, was the desire to fire a shot farther, harder, and more accurately. This, coupled with the increasing availability of gunpowder (and new propellants; see Anon. 1844, Buchanan 2006), would lead to the ability to fire larger and larger shots. Two areas of development that contributed to this goal that are not covered in this chapter are developments in gunpowder manufacture and testing and developments in ammunition, notably the shift from shot to shell that came in just before the Civil War. The former matter, however, does bear on cannon design, for it was just before the turn of the nineteenth century that experimenters began to get a handle on reliably quantifying the strength of gunpowder. By midcentury, military and university laboratories were investigating how to accurately measure gunpowder's effects (Woodbridge 1856; Kempers 1998:42–64). By the early nineteenth century, quantitative measures of the pressures inside a cannon began to be made—inaccurately at first, such as by Count Rumford who vastly overestimated them (Thompson 1781)—that then raised the question of how best to resist those pressures and transfer them to the projectile.

Consequently, in America from about 1830 onward the key objective among ordnance officers and designers was to try to make a larger gun that was both stronger and lighter. Strictly speaking, this has been a goal ever since the beginning of ordnance, but it was only after 1800 that this became a widespread and much discussed rationale for new ordnance designs. For example, Stephen Demainbray (1754:10, reproduced in Morton and Wess 1993:130) gave lectures on motion in the mid-eighteenth century in which he spoke of "new schemes of forges and other lighter cannons proposed for shipping, shewing what advantages and disadvantages may attend these modern improvements." All designers understood that a stronger gun, and specifically a stronger breech section, was necessary, though how to achieve that end was debated. Numerous approaches to solving the problem were proposed in the first half of the nineteenth century, and the fact that no one method was uniformly accepted shows how intractable a problem it was (potentially with many solutions, none necessarily better than any other). Larger guns uniformly went over to cast iron, while brass remained in use for some smaller field ordnance in the decades up to the Civil War. Inventors also deployed iron in new ways, mixing cast and wrought iron, introducing wire-wound guns, and beginning to experiment with steel for some parts of guns.

In retrospect it appears that the major bottleneck in creating a lighter and stronger gun was the casting technology of the day, but as few inventors could address that foundry art, they tried to work within existing manufacturing constraints. It had been long recognized that casting a cannon solid and boring out the chase, rather than casting with a core, resulted in a stronger mass of metal. In this Americans seem to have held their ordnance in high repute: a federal report (Seybert and Eustis 1811:303) claimed that by 1811 American foundries had "arrived at perfection." They noted that while "the art of boring cannon is, in many places of Europe, deemed a secret of great importance . . . [in] the United States, this process is so well understood, that an inspector of our artillery has declared to the world [that] he was never compelled to reject a gun on account of a defect in the bore." The difficulty, though, was that boring out a chase from the solid was difficult, slow, and expensive. Manufacturers would have been quite happy to find a way to cast around a core again, as had been done in the early history of artillery.

In 1900, a review (Taussig 1900:487) of the iron industry in America could say that just after the Civil War,

> the blast furnaces and iron works of the United States were behind
> Great Britain in their technology. Matters went much by rule of
> thumb. The ore and coal and flux were dumped into the furnace,
> and the product marketed as it chanced to turn out.

The reviewer implied that the situation was worse before the Civil War, though when one considers the specialized field of ordnance casting, a less-stark picture emerges. In a sense, Taussig had it half right: iron making certainly was rule of thumb, but that does not mean there was no development. Iron furnaces had retained essentially the same shape for many hundreds of years, but by the nineteenth century, and perhaps atypically in America, their shape began to transform. Furnace masters dispensed with centuries of accepted wisdom that tried to optimize the furnace to its ore. In 1854 the Englishman Frederick Overman (1854:150–151) made special note of the furnace at the West Point Foundry, one of the leading ordnance producers for the country at the time, noting that it was 25% taller than a contemporaneous Pennsylvania charcoal furnace (40 versus 32 feet with almost identical boshes: 9 feet versus 9 feet, 6 inches) and consequently could extract iron of "superior quality, very fusible and uniform," using about three-fourths the charcoal of the Pennsylvania furnace.

Ordnance manufacture also was at the peak of its art, as every single product was tested, some samples to destruction, and the casting bell and sprues from the guns were often tested for the properties of the metal. Qualitative records of the color, fracture, and structure of the iron were made for each cannon inspected, noting where it seemed too porous or too granular. Army and navy inspectors made note of what sources of iron went into the mixtures of gunmetal. It was clearly known that the pig iron made of ores from different mines, even two quite close together, had differing qualities. Engineers and foundrymen had always known that the strength of their metal was important in the success of their pieces, and by the early nineteenth century they were able to put numbers to this craft knowledge and demonstrate the superiority of ordnance metal. Thomas Telford, for example, had measured the crushing resistance of common gray cast iron at about 140,000 pounds per square inch and that of gunmetal to over two-and-a-half times that, 350,000 pounds per square inch (Wilson 1848).

Ordnance officers were also well aware of how cast iron cooled and the grain structures it developed. Army officers began speaking of the crystallization and grain of iron. Lt. John Gibbon reported to the Secretary of War in 1858 that he had just overseen the casting of an experimental 8-inch Columbiad (a style of gun; see Lewis 1963) at the West Point Foundry and noted that cast iron was one of the "crystallyine [sic] bodies [that] in passing from a liquid to a solid state arrange the crystal particles in directions normal [perpendicular] to the cooling surface" Gibbon (1858:1). It had for a long time been clear that the "re-entering angles" in a cannon's form were responsible for weak points, and Gibbon had noticed that what today would be called stress concentrations were causing the guns to fail with "remarkable uniformity." He surmised, however, that the failure was caused by the casting process, which affected grain structure. After a short didactic lesson (Figure 3.3) to the Secretary of War, Gibbon (1958) showed how sharp angles caused fracturable grain boundaries in the cannon and enclosed a large-scale drawing comparing the longitudinal cross sections of contemporaneous cannon. He proposed a process of annealing to reorient the grain structure and, to make his point clear, made an analogy to how ice forms and then recrystallizes over the late winter. Thus, while the modern science of metallographic analysis began with work by scientists such as Henry Clifton Sorby in the 1860s, army officers and ordnance foundries on the eve of the Civil War were paving the way in their need for strong, safe ordnance (Smith 1988).

There was, however, a problem (among others): James Renwick, a professor of natural philosophy at Columbia University in New York, noted that although cast iron had largely replaced brass in all large ordnance, "it has not been found

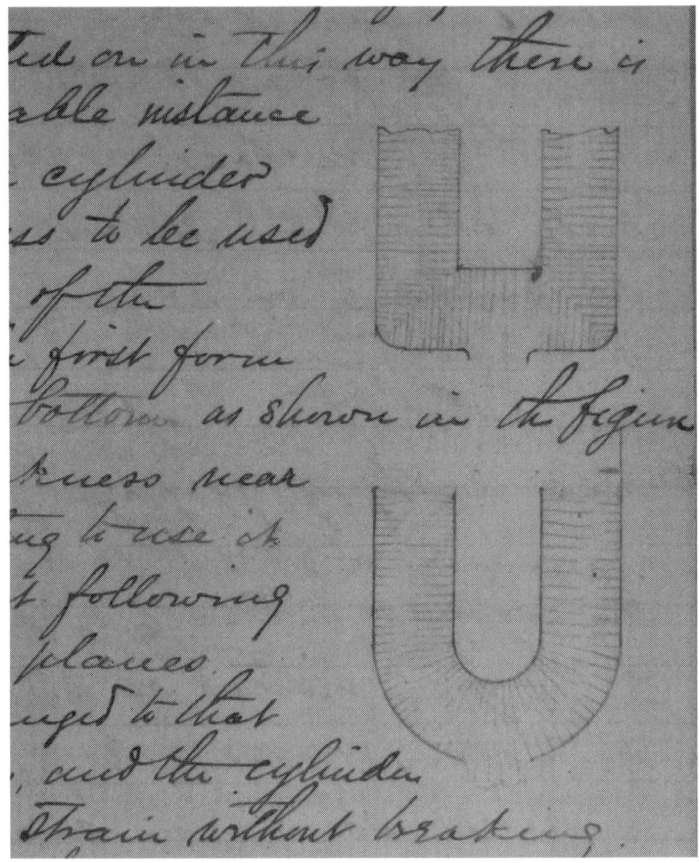

Figure 3.3. Detail of cast iron grain diagrams in a letter from Lt. John Gibbon to Secretary of War John B. Floyd, 6 December 1848 (Gibbon 1858, fol. 2). NARA RG156/E994 ('Special File'), box 1, no. 131. (Document photo by author.)

practicable, generally speaking, to use cast-iron in the lighter species of ordnance (field pieces)" (Renwick 1832: 262). Renwick and scientists of the day were aware of inverse scale effects—that a smaller version of a shape will be proportionally stronger—but there was a curious factor in cast iron that overrode this: "It is found that articles made from the same cast-iron, and drawn from the same charge of a furnace, will be weaker in proportion as they are more rapidly cooled." Therefore, smaller field pieces, having less mass and therefore less heat capacity, cooled faster and thus ended up proportionally *weaker* than larger guns of the same form. (Today we know that this is a function of grain growth and the phase in which the iron and carbon are locked upon crystallization.) Renwick nonetheless claimed that Swedish and American iron, made with charcoal, was more than sufficient to make small field pieces and that such pieces would in fact weigh less than brass

STEVEN A. WALTON

pieces of the same caliber and length (Renwick 1832:262; Rumford 1870–1875, vol. 1:183–184). Peter Barlow, a mathematician working at the Royal Military Academy at Woolwich in England (which was also the site of the Royal Brass Foundry for ordnance) also published an important paper at this time that would come to be seen as seminal in understanding the strength of cannon under pressure (Barlow 1836; see Whildin 1860:11–19), although by the later 1850s the British cannon inventor, Capt. Theophilus A. Blakely (1859:313), wrote that Barlow's paper had been "utterly neglected" at the time of its publication.

There were also questions about casting orientation. Scholars of early artillery argue that a great breakthrough in cannon strength came when founders inverted the guns and began to cast them muzzle up (Guilmartin 2003), and in the early nineteenth century some pressure experiments on cast iron that had been cast horizontally versus vertically—the latter providing a higher consolidation pressure on the breech from the column of molten iron—showed that iron cast under higher pressures had higher yield strengths (Wilson 1848, vol. 3:938–939; see also Rennie 1818 and 1819). Foundrymen also had differing opinions about whether using cold or hot air in the blast furnace during the primary smelting produced better iron: hot blast was more fuel efficient and probably did give better iron, yet the military adamantly mandated cold blast (Gibbon 1863:92) and, in inspection reports of cannon cast at the West Point Foundry between 1834 and 1840, even blamed an increase of second- and third-class ordnance on the use of hot-blast iron by the foundry (Walbach 1846:180).

Finally, beginning in 1845, after more than a decade of concerns about casting flaws and the merits of using hot-blast versus cold-blast casting, Capt. Thomas Jefferson Rodman redesigned the so-called Columbiad guns (an indistinct term at the time, though it seems to have referred to chambered guns; see Lewis 1963 and 1964) and patented a water-cooled core for casting the gun that forced the metal around the bore to solidify first (rather than last as it would have in the normal casting method), thereby improving the strength of very large and thick castings (Figure 3.4). His invention then spurred the civilian William Wade and other army ordnance officers to a whole series of experiments in the late 1840s and early 1850s on the effect of numerous variables, including chilling, on the properties of cannon metals (Ordnance Department 1856). Wade was a retired major in the Ordnance Department and former ordnance inspector at Fort Pitt Foundry in Pittsburgh who later became a cannon founder himself in the Pittsburgh firm of McClurg, Wade, and Company. As the "competent and skillful ordnance founder" (Poinsett 1841:2) asked for by the Secretary of War, he had also been a member of the War Department-sponsored trip to European cannon foundries in

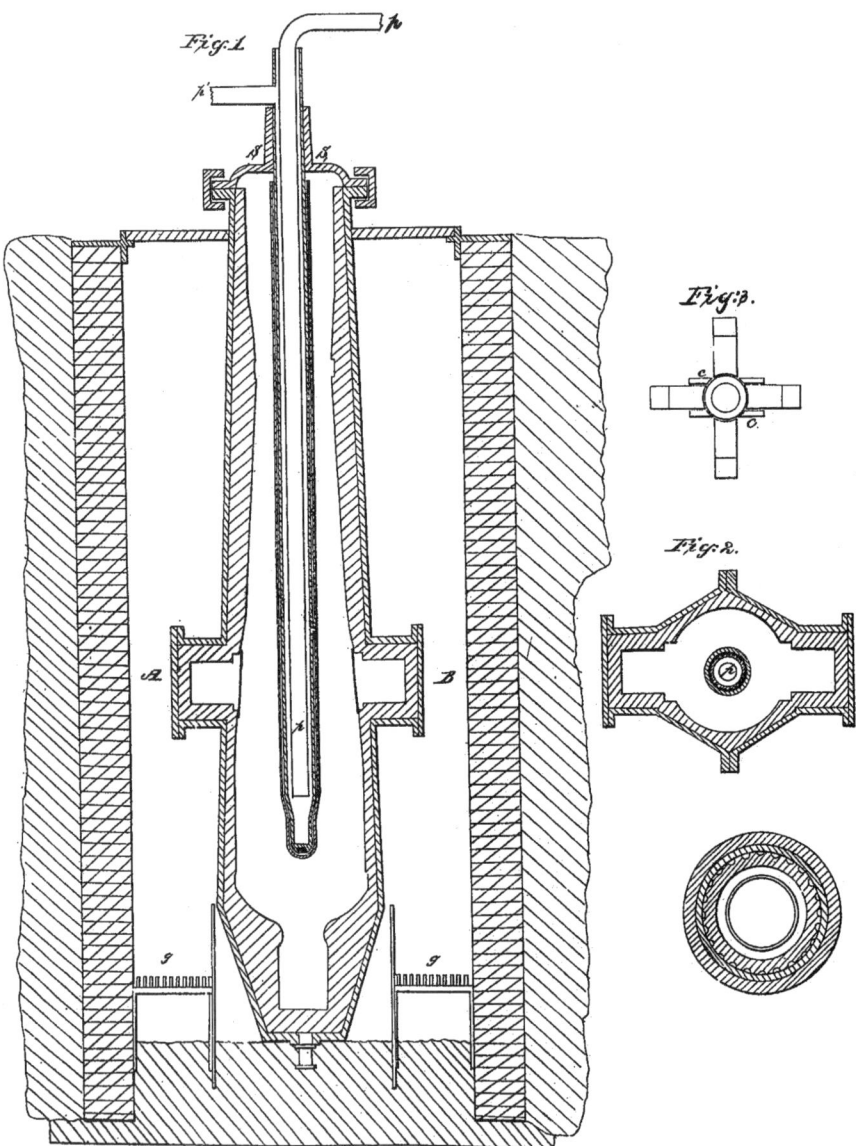

Figure 3.4. Diagram showing improvements in casting ordnance by Thomas J. Rodman, patented 14 August 1847. U.S. Patent and Trademark Office, Patent no. 5,236.

1840, along with Maj. Rufus L. "R.L." Baker, Capt. Alfred Mordecai, and Capt. Benjamin Huger. That commission brought many of the issues Wade would later investigate to the attention of the U.S. military. With Wade later casting cannon for the Union and Huger becoming a Confederate general and inspector of artillery and ordnance, the commission reflects how such knowledge fed into both sides' ordnance corps during the Civil War. (Baker spent his career in ordnance and was long assigned to Fort Monroe but had retired by the time the war broke out, and Mordecai resigned his commission rather than fight for either side.)

Thus it was that one of the major revolutions in artillery casting came not from a redesign of the shape of cannon (as casting over forging in the earliest years of artillery had done), nor from a change in the metallurgy of the raw materials (as the replacement if cast iron for cast bronze had done), but from a new method of casting cannon breech down with a hollow core and a core outfitted in such a way to fundamentally alter how the iron solidified. The Rodman method was used for his characteristically shaped guns, which we will consider shortly, but also for many other iron ordnance models cast for the U.S. military in the four major ordnance foundries after 1847.

SHAPED GUNS: MATCHING THE PRESSURES

In ordnance there was a continuous twofold demand: on one hand the military wanted a strong cannon that could take larger and larger gunpowder charges without rupturing, and on the other, it wanted to minimize the amount of material used so that the cannons would be both cheaper to manufacture (a concern of the treasury and for the foundries) and lighter to maneuver (a concern for the military). These demands led a number of artillerists to propose solutions to the problem.

What was just arriving was the birth of the field we now know of as strength of materials. Although the field can reasonably trace its origins to Galileo's *Two New Sciences* (1642), it would not be too far a stretch to say that not much had happened between then and the nineteenth century. The first book of modern metallurgy was published in the later 1860s, classical descriptions of strength of materials date to the 1870s and 1880s, and these descriptions were only codified in the early twentieth century (Todhunter and Pearson 1886:93; Timoshenko 1953). Stephen Timoshenko actively developed and formulated many of our modern conceptions for strength of materials between the world wars.

In the first third of the nineteenth century, a number of pioneers of the modern understanding of materials' strength began their research by finding ways to measure and to quantify stress (see Rogers and Roberts [1838] for the first

American textbook on the subject, though Mahan [1837] also touched on it). First they developed robust machines and measurement devices that could break iron and steel and measure stress and strain rates and breaking strengths, though it would not be until the 1840s that most U.S. ordnance foundries had testing machines installed (e.g., Dearborn 1845). At the same time, engineers and scientists began developing rudimentary closed, analytical frameworks—ideally, algebraic in form and founded on first principles of elasticity, stress, and strain—that could explain and predict behavior. It was this latter step that would eventually become very attractive to ordnance inventors as they sought ways to make cannon strong *and* light (Layton 1971:570–572).

As early as the mid-eighteenth century some scientists and engineers, such as Petrus van Musschenbroek (deceased 1761) and Emiland-Marie Gauthey (deceased 1807), developed reliable tensile testing machines, and by the end of that century they had effectively described failure modes and limits for many materials. Many of the greats in the history of mathematics—Coloumb, Navier, Poisson, and Cauchy—turned their attention to problems of stress and strain in materials and had begun to make progress on abstractions of material behavior and strengths by about 1835. The first systematic examination specifically devoted to the practical use of cast iron was made by Thomas Tredgold in England in 1824 (his book went through five editions by 1860), but as Tredgold was principally interested in structural iron, he measured and reported its behavior in tension, compression, torsion, and bending but not in pressure vessels as could have been useful in thinking about artillery. In yet another example of the military technology leading domestic uses, he reported,

> I was *led* into this important inquiry by considering the proportions
> for cannon, and the common method of proving them. It appears
> from my experiments, that firing a certain number of times with
> the same quantity of powder would burst a cannon when the strain
> is above the elastic force of the material, though the effect of the
> first charge might not be sensible. (Tredgold 1860:3 [italics added
> for emphasis])

Thus, this "new age in the history of machines" (Tredgold 1860:3) was opened by trying to understand how elongating forces could push a material past its elastic regime (where elongation is directly proportional to the force) into a deformation regime. In the case of cast iron, that deformation regime is usually quite small, which means the material quickly entered the final regime, failure, and in the case at hand, the cannon burst.

STEVEN A. WALTON

The problem was that most of the analytical techniques available at the time needed to assume uniformity—of pressure, of shape—in order to make any reasonable calculations. Many of the theories of strength of materials had been developed for soil pressures in abutments, for example, in canals (see Mukerji 2009 for seventeenth-century precursors). Once measurements could be made, the theory was ready to be put to work. Indeed, many of these early scientists were interested in the question of cannon pressures and failures, though the physical measurement of pressures in cannon was beyond mechanical ability of the time. Benjamin Thompson (Count Rumford) had (over)estimated the pressure inside a cannon to be 100,000 atmospheres (about 10 gigapascals), but other experimentalists began trying to figure out how pressures could be accurately measured.

Many scientists who looked into the topic had studied at the École Polytechnique, a military school in France. Their works were of great interest to both Sylvanus Thayer and Dennis Hart Mahan at the U.S. Military Academy at West Point, and the European ideas began to enter the American engineering system there as early as the 1820s. There were even calls from the popular press to look into this matter. In 1832, the *American Quarterly Review* reviewed *Elements of Mechanics* (Anon. 1832) by Prof. James Renwick of Columbia University and, in passing, noted that Renwick "gives many useful hints in relation to practical gunnery, which we think are well worthy of the consideration of the constituted authorities of our government" (Anon. 1832:145). While the reviewer concentrated on Renwick's thoughts on exterior ballistics, those comments came in the context of concerns about the internal ballistics that cause cannon failure. Renwick said that in various situations "the strength of material of which the piece is formed, may not be sufficient to resist the accumulation of force, and bursting may be the consequence." He said that he himself had witnessed proofs gone wrong (likely at the West Point Foundry, as he was friends with the Kemble brothers who owned it and was also one of the early investors in the foundry) where "the balls made their way through the sides of the piece, and large portions of the wad remained [stuck] to the bore in front of them" (Anon. 1832:145).

Renwick (1832; compare with Lallemand and Renwick 1820) had devoted a chapter to the strength of materials (book 3, chap. 7) and another to projectiles (book 4, chap. 3). In that latter chapter, he examined contemporary researches on exterior ballistics, but also included a subsection (§268) on the "Inquiry of the best figure of a cannon." He began with the obvious observation that the breech must be larger since it is subject to the greatest strain and hypothesized that "the greatest effort is exerted by the expanding gas, at the point where the ball is lodged . . . [though] it would be difficult to reduce it to the test of mathematical analysis" (Renwick 1832:262–263). He nodded to Count Rumford, who had proposed a gun that was thickest at this point and "of beautiful proportions, swelling in a curve from the breech to the point assumed for the lodgment of the ball, and again contracting in a curve to the projection of the muzzle." This description certainly prefigures the Dahlgren or Rodman shapes, though Rumford was writing at the turn of the nineteenth century (Rumford's plan for this cannon—as yet untraced—presumably arose out of his artillery experiments for the Duke of Bavaria in the 1790s; see Rumford 1870–1875, vol. 1:173–190). Renwick claimed that the American navy 32-pounder and battering 18-pounder were a preferred shape of a cylinder from the breech to the trunnions and then a section of a cone to the swell of the muzzle. By comparison, he criticized the American navy 42-pounder (based on the proportions of the French battering gun) as being too heavy in the breech and too thin in the "lodgment of the ball" and proposed that it could be improved by redesign according to the form of the 32-pounder.

By the late 1830s, debates about the shape of guns had so thoroughly penetrated thinking about cannon that in at least one case a manufacturer tried to see if these ideas were correct. The Alger Foundry in South Boston (later known as the South Boston Foundry) had been producing brass and iron guns for the army and navy as well as state militias for a decade or so when in 1841 it exhibited some 6-pounder iron guns at the Third Massachusetts Charitable Mechanics' Association fair. These guns had 10 holes, spaced two caliber-lengths apart, bored along their length. Each hole was threaded and fitted with a tapped pistol barrel that could hold a cast steel bullet. When the cannon was fired and all but one of the holes was sealed with screws, the bullet from the pistoled hole was ejected vertically with a force proportional to the pressure in the barrel at the point at which the pistol barrel was inserted. That force was then measured by how many half-inch pine boards the bullet penetrated (a traditional form of measuring the strength of gunpowder; see Kempers 1998). Alger did indeed demonstrate that the peak pressure was in the lodgement near the breach, where the projectile sat in contact with the powder charge, and that it decreased proportionally with distance forward from the vent. Consequently, the Mechanics' Association (MCMA 1841:66) reported that the resultant gun had been made by a "more accurate distribution of its material, [obtaining] an equal strength with less material." In awarding it a silver medal, they described Alger's artillery as follows:

> The guns exhibited were without the usual reinforces, the swell
> of the muzzle and several of the mouldings. Their form is that
> of a cylinder united by its base to a truncated cone, the plane of

conjunction containing the point where the ball and the cartridge are in contact, and perpendicular to the piece. The bore being a hollow cylinder, the conic form of the exterior graduates the thickness of the metal.

The guns seem to have more or less replicated what Renwick had proposed about a decade before, though with a longer taper and now with some experimental backing. In the end Alger claimed they saved about 15% of the weight over a standard pattern 6-pounder gun. Although these guns were never adopted by the military, this same rough shape would soon appear in Rodman's cannon within five years and then later in Dahlgren's a decade after that. It is worth noting that the mathematical and analytical analysis of the internal pressures in guns continued throughout the nineteenth century (Longridge 1860).

Thomas J. Rodman may have invented a new casting process that gave stronger cast iron guns, whatever their form, but he also investigated how the shape of a gun might be revised so that it would be stronger by its very nature. Although the very earliest cannon were not tapered from the breech to the muzzle, by as early as the sixteenth century it was intuitively understood—perhaps from watching far more breeches than muzzles explode—that a gun experiences a higher pressure in the chamber where the gunpowder charge detonates than it does at the mouth where the ball exits. Consequently, cast guns were cast with proportionally thicker breeches. But thicker breeches meant heavier guns that were harder to cast and harder to transport as well as guns that were more expensive, so in the new nineteenth-century understanding of optimizing a machine's engineering efficiency (Alexander 2008), Rodman set out to *calculate* the optimal shape of a cannon to resist the internal pressures uniformly. His solution was to make the wall thickness directly proportional to internal pressure along the barrel.

Rodman developed an external profile for the gun that was, he argued, in proportion to those internal pressures and whose swelling silhouette more efficiently resisted those pressures during firing (Figure 3.5). In his method, as recorded by British military officers visiting from Canada early in the Civil War, the designer takes the desired caliber of the gun and sets the chase of the barrel to 10 caliber-lengths (150 inches for a 15-inch gun, e.g., as in Figure 3.5). The required thicknesses at the breech (*OP*) having been obtained through strength testing (or deduction) and the muzzle thickness (*QR*) being set to one-third of that thickness, Rodman then proposed that the *oblique thickness* (the slanted dashed lines) at various points from the muzzle to the breech remain constant. The shape of the gun was then evolved by setting *QS* = *OP* and extending those two lines until they

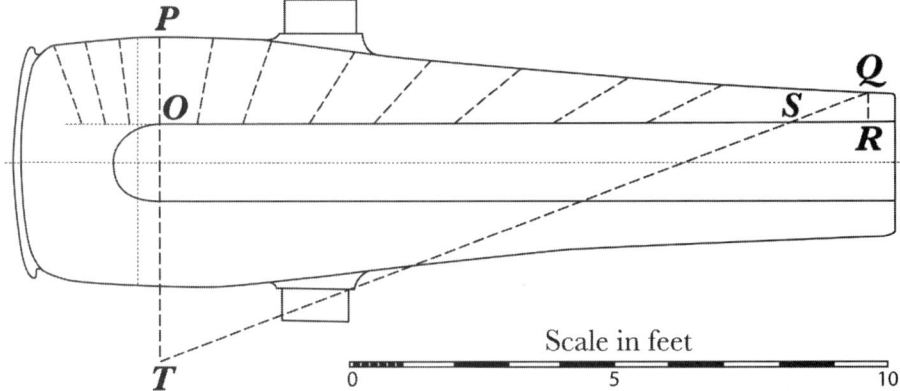

Figure 3.5. Graphical construction method of a cast-iron 15-inch Columbiad. The delineator has noted the construction method as follows: "Construction. O.P. = 16½ Inches. Q.R. = ⅓ of O.P. = 5½ Inches. O.R. = 10 Calibers = 150 Inches. [To] Produce P.O. make Q.S. = O.P. Produce Q.S. to meet P.O. produced in T. Then if line be drawn from T. intersecting O.R. and from the points of intersection distances be set off equal to O.P, these points will give the exterior line of the gun." Redrawn (with specific dimensions removed) from Report of Officers of the British Army on American Arms, 1862, pl. 4 [The National Archives (Kew, UK) WO 33/11, p. 731].

meet at *T*. Point *T* then becomes the center from which construction lines are rotated and the oblique thickness *OP* marked off for the entire length of the barrel, thus generating the silhouette. While the British visitors reported to their war office in Whitehall that they found his approach very attractive (though admittedly "peculiar"), the five pages of equations given by Rodman (1861:217–222) to define the shape of his cannon were far too "abstruse" (Williams 1862:28). They proposed streamlining them to an equation using two simple moduli for the proportion of the desired gun based on the gun's proportions and the square root of the difference of squares. In Rodman's method one can see the experimental understanding of pressure distribution in a cannon barrel merging with the desire to describe that barrel geometrically. Although a number of the steps still remained somewhat arbitrary or insufficiently justified (e.g., that the breech–muzzle thickness should be exactly a factor of 3 or that a constant oblique thickness was the same as the fall-off of pressure from breech to muzzle), his approach was fundamental in reforming the understanding of how one designs and describes the shape of a cast iron cannon.

On the navy side, Lt. (later R. Adm.) John A. Dahlgren also developed guns that had a swelling shape, and both he and Rodman were well versed in calculus and contemporaneous theories of pressures in a barrel. Dahlgren began working on the problem in the early 1850s (see Dahlgren 1856) and patented his system, where the "quantities of metal disposed in the different parts of the gun are

STEVEN A. WALTON

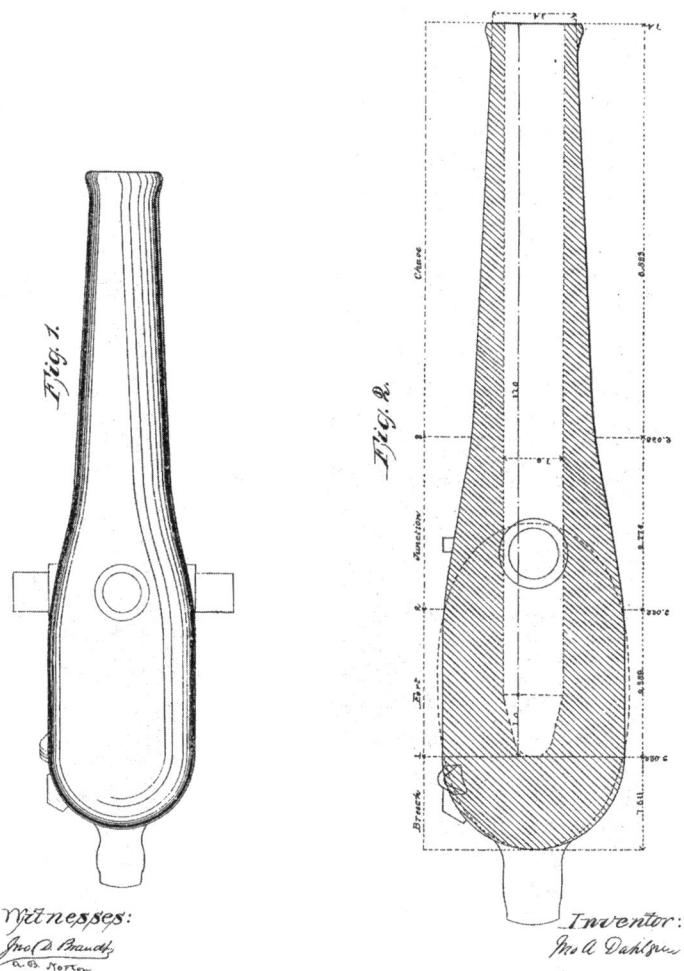

Figure 3.6. "Cast Iron Ordnance" diagram by John A. Dahlgren, patented 6 August 1861. U.S. Patent and Trademark Office, Patent no. 32,983.

proportionate, or nearly so, to the relative degrees of strain exerted by the force of the exploded charge at those parts respectively" (Dahlgren 1861:2) (Figure 3.6). Although he speaks of a "rule" for the shape of his gun (though at the same time admitting that he deviated "from strict theoretic proportions"), it is clear that the shape is nonanalytical (i.e., not one smooth, continuous curve that can be written as an equation or geometric rules, like Rodman's) and that he thought of the gun as being described by four separate functions or curves for each of its sections: the breech, fort, junction, and chase, as he called them. He mentions that if one wished to make a gun of a different caliber (the defining modulus of construction), then "the dimensions in calibers of all the other parts may be obtained by measuring the drawing" (Dahlgren 1861:1). The army would come to rely almost exclusively on Rodman guns to meet its requirement for large siege and costal defense guns and the navy brought Dahlgren guns into use aboard most of its ships. They became the norm for large-caliber shipboard ordnance by the Civil War. The thousands of Rodman and Dahlgren guns produced in the 1850s and 1860s became a mainstay of both sides during the Civil War (Dahlgren 1887).

There were also inventors who tried to work with fundamental metallurgical properties of iron in order to produce stronger cannon during this period. Norman Wiard, originally from Ontario, became an Inspector of Ordnance Stores by the Civil War and proposed a cannon cast from puddled wrought iron, which was a pre-Bessemer method of tempering the carbon content of cast iron into the range of carbon steel. Also aware that the shape of a cannon was in play, his cannon had a cylindrical rear half with a hemispherical breech, a midsection stepped just ahead of the trunnions, and a fore end tapered linearly to the muzzle. He made a number of both rifled and smoothbore 6- and 12-pounders, as well as a boat howitzer that were accepted by the Union army. Although the British commission dismissed the Wiard gun as an overly complex construction that did "not appear to have any advantage in enduring qualities, [and whose] principles . . . do not give promise of a strong gun" (Williams 1862:26), the guns did see some deployment by the Civil War. The British largely attributed this to political influence rather than any inherent advantage, and only a few dozen were ever made and largely went out of favor after the war (Hazlett et al. 2004:163–167). For his part, Wiard continued to experiment with artillery and tried for years to re-interest the government in his designs (Wiard 1895).

BANDED GUNS:
CONSTRAINING THE PRESSURES

While various inventors were thrashing about with an emergent understanding of the strength of materials and designing guns at the very same time that their knowledge of the material properties of both the cannon materials and the internal ballistic pressures was itself developing, another set of inventors sought a different solution to the reverse salient problem of guns over-pressuring. From the earliest days of artillery construction, gun makers had employed hoops of wrought iron to band and reinforce the breech—and in the earliest days the entire barrel—to make sure that the cannon would withstand the pressure of firing. Though that solution largely disappeared by the seventeenth century, the fundamental concept of strengthening hoops would have been known to any blacksmith as an obvious solution to making a hollow structure stronger. Therefore, in the nineteenth century a second general solution to the problem of higher pressures within guns was to reinforce the breech with larger bands of strong iron—initially wrought iron but later steel—in such a way that although the cast iron barrel of the gun would not have alone been able to resist the pressure of the charge, the composite cast and wrought iron gun could. Such a system also allowed for the work of building guns to be broken down into smaller processes, each at least potentially with more quality control.

The earliest American attempts at this strategy were undertaken by Daniel Treadwell, the Rumford Professor at Harvard from 1834 to 1845. Having come up through the trades rather than a product of a college education or of the military, Treadwell emphasized the practical rather than theoretical basis of invention, and he was quite willing to disparage then-modern academics and philosophers for being "far too willing to follow the lead of Plato in despising 'servile subjects' connected to labor and the useful arts" (James 1992:62–63). His original connection to the military seems to have come though the installation of his innovative rope-making machines at the Charlestown Navy Yard in the 1830s. By 1840, Treadwell had also recognized the need for a strong but light iron gun. His solution was to make a cast iron tube and then reinforce the breach end with a band of wrought iron (Treadwell 1845, 1865). Fundamentally, these are the first of what will eventually be known as "built-up" guns, though that term is more often associated with the guns of the English inventors Blakely and Armstrong from the 1850s (Bastable 1992; Hall 2001). That this was a solution founded in ideas of the differential material properties of wrought and cast iron is obvious when one realizes that Treadwell went to great lengths to blend the wrought iron into the cast so that the cannon looked similar to a regular cast pattern 1845 cannon (Figure 3.7). By 1844, Treadwell had the army's ear, and guns were ordered, manufactured, and delivered for trials. In 1844 and 1845, Treadwell awaited the results anxiously; he founded the U.S. Steel Cannon Company (bankrolled by Francis Cabot Lowell) in Boston and also resigned from his Harvard Rumford professorship, citing too little time to devote to teaching.

Treadwell published a memorial in the mid-1850s on built-up guns and offered his guns for trial in France and in Great Britain. France tried them and

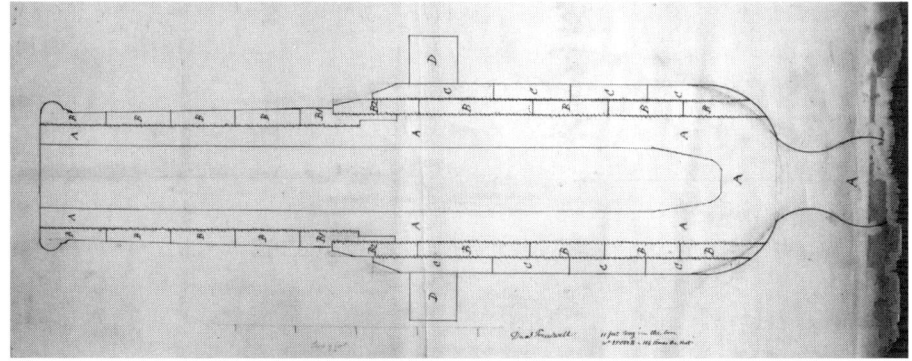

Figure 3.7. Drawing by Daniel Treadwell of a composite hooped gun, undated, but likely ca. 1850. Harvard University Archives, Cambridge, Mass., HUG1847.505pF, undated folder.

issued a noncommittal report. Britain stalled, and even though Treadwell moved to London to promote his methods (just at the very time that Blakely and Armstrong, too, were starting to build built-up guns; see Stoney 1870b), he failed to interest the Ordnance Select Committee, and they declined to try his guns. Although his grand cannon business was in the end a commercial failure, he himself remained wealthy through his printing and other patents.

It is clear that the idea of shrinking a band on the breech of a cannon was being discussed in many halls of the military (not to mention by civilians) in the 1850s. In one case in 1860, Maj. Thomas J. Rodman was asked to comment on the system of built-up guns that was at that time attached to a proposal by Maj. Irvin McDowell (shortly to be defeated at fist Manassas). Rodman (1860) thought that if it were done *accurately* (his emphasis), such guns would be stronger and lighter, but in estimating that he speaks in terms of what sounds like material science (Rodman 1860:1):

> The maximum resistance which this method is capable of securing,
> would be obtained by shrinking [on] . . . a number of layers of very
> thin concentric rings one upon another in such manner and which
> such strain or tension as would cause them & the core, all, under
> the bursting effort of the charge, to be brought to their *breaking
> strains* simultaneously.

Rodman proceeded by logic, though not by mathematics, to argue that a thicker hoop, or a fewer number of hoops, would be weaker because when the rupture happens from within, the breaking strain will first be reached by any inside face of a hoop and then once past that breaking strain, the crack would propagate

STEVEN A. WALTON

outward through the entire ring. With thinner rings, any single crack could not grow as long as the next ring held. Rodman also had a clear understanding of contemporaneous knowledge about the relation of longitudinal and circumferential stress (hoop stress, as it is now called), though he related it in terms of resistance and indicated that the former would be more than half the hoop stress (modern understanding has longitudinal strain as a function of the pressure and the proportion of internal diameter to wall thickness). In the end, Rodman was concerned about using old guns and rehooping them for strength but did think it advisable to first test the method with "new cores, & with *steel* instead of wrought iron rings" (Rodman 1860:3, emphasis his).

In terms of widely used built-up cannon of the Civil War, no inventor is better known than Robert Parker Parrott. An 1824 graduate of West Point, Parrott was first assigned to the artillery and then served, briefly, as captain in the Ordnance Corps. Like Wade in Pittsburgh, he resigned his commission and took over the superintendence of the West Point Foundry in Cold Spring, New York, across the river from United States Military Academy at West Point, where he had once been an inspector (he also married Mary Kemble, sister of the foundry's owners). Parrott ran a foundry that was doing a booming business with both commercial and military orders for cast iron goods, ranging from cannon to sugar presses and steam engines. His solution to the problem of gun pressures was to put a thick wrought iron band around the breech of a cast iron tube, achieving by a simple composite structure what Rodman and Dahlgren tried to do in the casting alone. The system was a great success and evolved into nearly a dozen variations in size from the 3.67-inch 20-pounder field piece, the 6.4-inch 100-pounder used in fortifications and in naval situations, and the massive 300-pounder siege guns that saw service for the army and navy during the Civil War (Figure 3.8). And as imitation is the sincerest form of flattery, very quickly the Confederacy began banding any old ordnance it could, and the naval officer John Mercer Brooke, now the Confederate in charge of the conversion of the USS *Merrimack* into the CSS *Virginia*, designed the Brooke rifle along the same principles (Brooke 2002).

Parrott guns went into production immediately as the war began, though it seems the idea of the Parrott gun may have been developed as early as 1842 at the West Point Foundry, although not put to trial until later. In a letter to Col. George Talcott, head of the Army Ordnance Department, foundry owner Gouverneur Kemble wrote,

> We are not fond of vague experiments, but we think that your field artillery might be lightened, if useful, below the weight of brass

Figure 3.8. Top: A 100-pounder Parrott naval cannon, Xenia, Ohio. (Photo by author.) Bottom: A 300-pounder Parrott with burst muzzle, Morris Island, South Carolina. (Photo courtesy Library of Congress, Prints and Photographs Division, Washington, D.C., LC-DIG-cwpb-04726.)

guns, and at the same time cheapened, by hooping a cast iron gun behind the reinforce with wrought iron hoops. We should turn the gun, and bore the hoops, to a diameter something less than the piece, and after expanding them by heating, shrink them on. If you deem it of sufficient importance to subject a gun of this description

STEVEN A. WALTON

to the proper tests and trials, we would make one at our cost and risk, after the model of your light 6p^r of 1838, or construct a gun of such number of calibers and weight as you might deem of the greatest advantage in service. (Kemble 1842:2–3)

Kemble, who had just returned to Cold Spring after two terms as a U.S. Representative during the Van Buren administration and was a close personal friend of many West Pointers, certainly had Talcott's ear. Talcott referred the letter up to the Secretary of War, John C. Spencer, and within two weeks Talcott had given the foundry an order for $25,000 in coast howitzers to keep them in business; Spencer added an endorsement on the letter: "I see no objection to his making a gun at his cost hooped as proposed." It is unknown whether the West Point Foundry did ever try those guns at their own cost at that time.

During the Civil War, Parrott managed to build up a strong mythology of his own heroic independent inventor status. As reported in many newspaper outlets of the period, it went something like this (from the *Chicago Tribune* [Anon. 1863:3] discussing the huge guns used at the siege of Charleston in 1863):

The larger a gun having a rifle bore is made the steadier and truer the ball or shell is sent—indeed the accuracy of a rifle is attained. Mr. R.P. Parrott, the inventor of these guns, commenced making them in 1856, at his own expense, continuing his experiments without aid from the Government, till the rebellion broke out. He then began to make ten pounders, and has now advanced from that small beginning to three-hundred pounders. He is at present engaged in a five-hundred pounder, and will try a two-thousand pounder. Over twenty-five hundred of these guns have been made by Mr. Parrott, who furnishes them at less cost than the Government can make them at its own foundries; indeed, supplying them at a trifle above cost, depending upon shells, which he furnishes to the Government, for his profits.

Parrot was at the time turning about three dozen guns a week on average, and though the *Chicago Tribune* article suffers from considerable hyperbole, it does capture the place of Parrott's invention in the history of American artillery.

Although the Union placed immediate and very large orders for Parrott guns at the opening of the war, there were some strong voices ("the most scientific officers in the United States"; Williams 1862:22) against them, even before some began demonstrating ruptures. More interesting is that the arguments—eventually borne out by the types of failures that did occur—were fundamentally material in their conception. The British commission in 1862 relayed to Whitehall that

> the cheapness of manufacture for this gun is a great argument in its favour, joined with sufficient strength to do the work for a considerable time. But although the exigencies of war have been the means of bringing it into such extensive use, yet among the ordnance corps it has few supporters; their opinion is against the combination of the two metals, from the fact that their tensile, expansile, and elastic properties being different; and although many of these guns have stood without apparent injury over 1,000 rounds, yet the moment may be looked for in which the cohesive force of the wrought iron will be destroyed, from continued vibration of the particles, and the tension of safety removed from the cast iron will cause its destruction. . . . The practical success of the gun is, however, a strong opponent to the theory, and a careful investigation by experiment is the only mode to arrive at the truth. (Williams 1862:22)

It is worth noting how these military officers were entirely conversant in the terminologies—tensile, expansile, and elastic properties, forces, and vibration—that would have been used by physicists of the day in trying to describe material behavior.

Parrott gave his first indication to the War Department of his intention to manufacture hoop-reinforced guns on 10 April 1861, two days *before* General Beauregard fired on Fort Sumter. Within months, Treadwell appears to have gone livid, or at least as livid as one can look though manuscript letters to formal officials. By the end of the year, he had sent three reports to the Ordnance Department and had had Lowell write a letter to their Massachusetts congressman claiming originality (Treadwell 1861a, b, c). Shortly thereafter, he published another memorial, this time claiming priority and also sent privately printed memorials to congress, as well as initiating a lawsuit against Parrott based upon his 1855 patent. Parrott claimed his investigations went back to 1856 and that he had had no knowledge of Treadwell's attempts to get the army interested in his cannon in the late 1840s and early 1850s. Ultimately, Treadwell's patent suit against Parrott was dismissed in 1866 (for the litigation, see patent cases 5 Blatch. 370, 3 Fisher 124, and 24 F. Cas. 154; see also Anon. 1866:1).

It is clear, then, that the period from 1830 to 1860 was a period of intense experimentation, variation, and diversification of artillery types in America, and this says nothing of the myriad European developments in the same period. Many inventors labored to produce reliable, light, and powerful artillery pieces, They varied in size, form, composition, and manufacturing method, and for every Parrott, Armstrong, or Krupp, whose eponymous ordnance won each an international reputation, there were a dozen Wiards, Blakelys, Jameses, and Stocktons whose names remained obscure (Hazlett et al. 2004). At the end of this period, steel manufacture was revolutionized by the Bessemer process (patented in England in 1855), which allowed huge quantities of cast steel to be made in foundries, and this then led to a whole separate material revolution in ordnance after the Civil War.

The revolutions in ordnance in America before the war were driven by a series of interlocking questions and methods, including questions about pressures inside barrels, the advent of instrumental and analytical techniques to measure these pressures, the arrival of theoretical understandings to describe the stresses and strains in a barrel (often mathematically, if even only qualitatively at first), and the continuing development of the nascent science of metallography and materials science. The 1863 *Treadwell v. Parrott* lawsuit brought many of the fundamental questions about these developments to the fore, asking questions such as, What are the fundamental distinctions between material and form? How does manufacturing process affect the nature of an object? How does one relate a material idea to a performance reality?

In small arms there had been some notable developments well before the Civil War, the most famous of which were the armory system of interchangeable parts and Samuel Colt's revolver (Smith 1980; Hounshell 1984). Before the Civil War, the fundamental terms of how large guns behaved were imperfectly worked out, but the war forced numerous types of old and new guns into the field out of necessity. It was not, however, until after the Civil War that American inventors really blossomed in term of military inventions that had staying power (Bartlett 1883) and the sciences of strength of materials and metallurgy matured to the point that successful predictive invention in ordnance could take place. Even then, in terms of heavy ordnance, the European modes of manufacturing built-up guns worked their way to America and, even before 1880, ordnance development plateaued in terms of method—in size, though, guns kept growing and growing.

Note

1. Tennyson (1870) had watched the bloody and futile charge at Balaclava from a distance during the Crimean War. His original 1854 poem, "The Charge of the Light Brigade," reads: "Cannon to right of them, Cannon to left of them, / Cannon in front of them Volley'd and thunder'd; / Storm'd at with shot and shell, Boldly they rode and well, / Into the jaws of Death, Into the mouth of hell Rode the six hundred." For the connection between the Mexican–American War and the Civil War, see Doughtery (2007).

References

Abbreviations

NARA   National Archives and Record Administration, Washington, D.C.

RG   Record Group

TNA   The National Archives, Kew [London], UK

Alexander, Jennifer Karns. 2008. *The Mantra of Efficiency: From Waterwheel to Social Control*. Baltimore: Johns Hopkins University Press.

Anonymous. 1832. Renwick's Mechanics. *American Quarterly Review*, 11, no. 21:120–153.

———. 1844. On Explosions and Explosive Compounds. *The Living Age*, 33, no. 25:43–46.

———. 1863. "The Siege of Charleston. Why the Shelling of the City Was Discontinued. Greek Fire and the Parrot [*sic*] Guns." *Chicago Tribune,* 10 September 1863, p. 3.

———. 1866. "From New York." *Chicago Tribune,* 8 December 1866, p. 1.

———. 1870. The Construction of Heavy Artillery. *Nature*, 3(24 Nov.):69–73, (15 Dec.):128–132.

Army, Thomas F., Jr. 2016. *Engineering Victory: How Technology Won the Civil War*. Baltimore: Johns Hopkins University Press.

Barlow, Peter. 1836. On the Force Excited by Hydraulic Pressure in a Bramah Press; The Resisting Power of the Cylinder, and Rules for Computing the Thickness of Metal for Presses of Various Powers and Dimensions. *Transactions of the Institution of Civil Engineers*, 1:133–139.

Bartlett, Wallace A. 1883. *Some Weapons of War as Improved by Recent American Inventors*. Washington, D.C.: National Republican Press.

Bastable, Marshall J. 1992. From Breechloaders to Monster Guns: Sir William Armstrong and the Invention of Modern Artillery, 1854–1880. *Technology and Culture*, 33, no. 2(Apr.):213–247.

Blakely, Theophilus Alexander. 1859. A Mode of Constructing Cannon, Whereby the Strain Produced by Firing Is Distributed throughout the Mass of Metal. *Journal of the Royal United Service Institution*, 2:304–337.

Brooke, John Mercer. 2002. *Ironclads and Big Guns of the Confederacy: The Journal and Letters of John M. Brooke*. Columbia: University of South Carolina Press.

Bruce, Robert V. 1956. *Lincoln and the Tools of War*. Indianapolis, Ind.: Bobbs-Merrill.

Buchanan, Brenda S., ed. 2006. *Gunpowder, Explosives and the State: A Technological History*. Aldershot, UK: Ashgate.

Dahlgren, Charles Bunker. 1887. *The Dahlgren Shell-Gun and Its Services during the Late Civil War*. Trenton, N.J.

Dahlgren, John A. 1856. *Shells and Shell-Guns*. Philadelphia: King and Baird.

Dearborn, Lieut. A.H. 1845. Letter to Col. George Talcott, 11 July 1845. NARA RG 156/E987, box 28, letter no. 66 for 1845.

Demainbray, [Stephen Charles] Triboudet. 1754. *The Syllabus of the Course of Natural and Experimental Philosophy*. London.

Dougherty, Kevin. 2007. *Civil War Leadership and Mexican War Experience*. Jackson: University Press of Mississippi.

Gibbon, John. 1858. Letter to Secretary of War, John B. Floyd, Report IN-1a-131, 6 December 1858. NARA RG156/E994 ('Special File'), box 1, no. 131.

———. 1863. *Artillerist's Manual, Compiled from Various Sources, and Adapted to the Service of the United States*. 2nd ed. New York: D. Van Nostrand.

Guilmartin, John Francis, Jr. 2003. *Gunpowder and Galleys: Changing Technology and Mediterranean Warfare at Sea in the Sixteenth Century*. 2nd ed. Annapolis, Md.: Naval Institute Press.

Hall, Nicholas. 2001. Theophilus Alexander Blakely and the Revolution in Victorian Gun Design. *Royal Armouries Yearbook*, 6:134–149.

Hazlett, James C., Edwin Olmstead, and M. Hume Parks. 2004. *Field Artillery Weapons of the Civil War*. 2nd ed. Urbana-Champagne: University of Illinois Press.

Hounshell, David A. 1984. *From the American System to Mass Production, 1800–1932: The Development of Manufacturing Technology in the United States*. Baltimore: Johns Hopkins University Press.

Hunt, Henry J. 1891. Our Experience in Artillery Administration. *Journal of the Military Service Institution*, 12:197–224. (Pp. 214–223 extracted in Tibdall 2011:398–407.)

James, Mary Ann. 1992. "Engineering and Environment for Change: Bigelow, Pierce, and Early Nineteenth-Century Practical Education at Harvard." In *Science at Harvard University: Historical Perspectives*, ed. Clark A. Elliott, pp. 55–75. Bethlehem, Pa.: Lehigh University Press.

Kemble, Gouverneur. 1842. Letter to Col. George Talcott, 2 Nov 1842. NARA RG156/E192/no.76.

Kempers, R.T.W. 1998. *Eprouvettes: A Comprehensive Study of Early Devices for the Testing of Gunpowder*. Leeds, UK: Royal Armouries Museum.

Lallemand, Henri, and James Renwick. 1820. *A Treatise on Artillery*. New York: C.S. Van Winkle.

Layton, Edwin. 1971. Mirror-Image Twins: The Communities of Science and Technology in 19th-Century America. *Technology and Culture*, 12, no. 4(Oct.):562–580.

Lewis, Emanuel Raymond. 1963. *The Ambiguous Columbiads: Historical Notes on the Origins of a Weapon and Its Name*. N.p.: U.S. Coast Artillery Historical Association.

———. 1964. The Ambiguous Columbiads. *Military Affairs*, 28, no. 3(July):111–122.

Longridge, James Atkinson. 1860. On the Construction of Artillery, and Other Vessels, to Resist Great Internal Pressure. *Minutes of the Proceedings of the Institution of Civil Engineers,* 19:283–337.

Mahan, Dennis Hart. 1837. *An Elementary Course of Civil Engineering for the Use of the Cadets of the United States' Military Academy*. New York: Wiley and Putnam.

Mallet, Robert. 1856. On the Physical Conditions Involved in the Construction of Artillery, and on Some Hitherto Unexplained Causes of the Destruction of Cannon in Service. *Transactions of the Royal Irish Academy*, 23:141–436.

Manucy, Albert C. 1949. *Artillery through the Ages: A Short Illustrated History of Cannon, Emphasizing Types Used in America*. Washington, D.C.: Government Printing Office.

MCMA (Massachusetts Charitable Mechanics Association). 1841. *The Third Exhibition of the Massachusetts Charitable Mechanics Association*. Boston: T.R. Marvin.

Morton, Alan Q., and J.A. Wess. 1993. *Public and Private Science: The King George III Collection*. Oxford, UK: Oxford University Press, with the Science Museum.

Mukerji, Chandra. 2009. *Impossible Engineering. Technology and Territoriality on the Canal du Midi*. Princeton, N.J.: Princeton University Press.

Ordnance Department (U.S. Army). 1856. *Reports of Experiments on the Strength and Other Properties of Metals for Cannon, with a Description of the Machines for Testing Metals, and of the Classification of Cannon in Service*. Philadelphia: Henry Carey Baird.

Overman, Frederick. 1854. *The Manufacture of Iron, in All its Various Branches*. Philadelphia: Henry C. Baird.

Poinsett, Joel Roberts. 1841. *Documents Relating to the Improvement of the System of Artillery*. Washington, D.C.: Blair & Rives. (Document from 26th Congress, 2nd Session, U.S. Senate doc. no. 229.)

Rennie, George. 1818. Account of Experiments Made on the Strength of Materials. *Philosophical Transactions of the Royal Society of London*, 108:118–136.

———. 1819. Account of Experiments Made on the Strength of Materials. *Philosophical Magazine*, 1st ser., 53, no. 251:161–175.

Renwick, James. 1832. *Elements of Mechanics*. Philadelphia: Carey and Lea.

Rodman, Thomas J. 1860. "Views of Capt. Rodman on the Method Proposed by Major McDowell for Strengthening Cannon," Report IN-1a-139, 9 January 1860. NARA RG156/E994 ('Special File'), box 1, no. 139.

———. 1861. *Reports of Experiments on the Properties of Metals for Cannon, and the Qualities of Cannon Powder*. Boston: C.H. Crosby.

Rogers, William Barton, and Verne L. Roberts. 1838. *An Elementary Treatise on the Strength of Materials*. Charlottesville, Va.: Tompkins and Noel.

Ross, Charles D. 2000. *Trial by Fire: Science, Technology, and the Civil War*. Shippensburg, Pa.: White Mane Books.

Rumford, Benjamin. 1870–1875. *The Complete Works of Count Rumford*. 4 vols. Boston: American Academy of Arts and Sciences.

Seybert, Adam, and William Eustis. 1811. "Cannon, Small Arms, and Other Munitions." In Select Committee on Manufacture of Cannon and Small Arms and the Munitions of War, 16 December 1811. *American State Papers: Military Affairs* no. 105, ASP016, pp. 303–307.

Smith, Cyril Stanley. 1988. *A History of Metallography: The Development of Ideas on the Structure of Metals before 1890*. Cambridge, Mass.: MIT Press.

Smith, Merritt Roe. 1980. *Harpers Ferry Armory and the New Technology: The Challenge of Change*. Ithaca, N.Y.: Cornell University Press.

Stoney, Capt. F.S. 1870a. A Brief Historical Sketch of Our Rifled Ordnance from 1858 to 1868. *Minutes of Proceedings of the Royal Artillery Institution*, 6:89–122. Reprinted Huntingdon, UK: Ken Trotman, 2004.

———. 1870b. The Construction of Our Heavy Guns. *Minutes of Proceedings of the Royal Artillery Institution*, 6:406–432.

Swain, Craig. 2013. Ordnance Observations from the Field at Charleston. *To the Sound of Guns: Civil War Artillery, Battlefields, and Historical Markers.* http://markerhunter.wordpress.com/2013/10/07/barnwell-ordn-notes/ (accessed 11 June 2014).

Taussig, F.W. 1900. The Iron Industry of the United States. *Quarterly Journal of Economics,* 14:143–170, 475–508.

Tennyson, Alfred Lord. 1870. Charge of the Light Brigade. In *Poems of Alfred Tennyson.* Boston: J.E. Tilton.

Thompson, Benjamin [Count Rumford]. 1781. New Experiments upon Gun-Powder, with Occasional Observations and Practical Inferences; to Which Are Added, an Account of a New Method of Determining the Velocities of All Kinds of Military Projectiles, and the Description of a Very Accurate Eprouvette for Gun-Powder. *Philosophical Transactions of the Royal Society of London,* 71:229–328.

Tidball, John C. 2011. *The Artillery Service in the War of the Rebellion, 1861–65.* Yardley, Pa.: Westholme.

Timoshenko, Stephen 1953. *History of Strength of Materials.* New York: McGraw-Hill.

Todhunter, I., and Karl Pearson. 1886–1893. *A History of the Theory of Elasticity and of the Strength of Materials, from Galilei to the Present Time.* Cambridge, UK: Cambridge University Press.

Treadwell, Daniel. 1845. *A Short Account of an Improved Cannon.* Cambridge, Mass.: Metcalf.

———. 1861a. "Papers in relation to the subject of hooped guns. Claims to be the originator of the idea, since acted upon in the Blakely and other guns." Report no. 189. Cambridge, 23 December 1861. NARA RG156/E192.

———. 1861b. "Francis C. Lowell writes to the Hon. Samuel Hooper calling attention to Prof. Treadwell's cannon." Report no. 193. Boston, 22 December 1861. NARA RG156/E192.

———. 1861c. "Treadwell sends papers relating to his invention." Report no. 194. Boston, 23 December 1861. NARA RG156/E192.

———. 1865. *Papers and Memoirs Concerning the Improvement of Cannon Published Between the Years 1845 and 1862.* Cambridge, Mass.: Harvard University Press.

Tredgold, Thomas. 1860. *Practical Essay on the Strength of Cast Iron, and Other Metals.* 5th ed. London: John Weale. [1st ed. published 1824.]

Walbach, Lt. L.A.B. 1846. "Examination of Iron Ordnance." In *Report of the Chief of Ordnance [for 1846],* Lt. Col. George Talcott, pp. 177–189. Washington, D.C.: Government Printing Office.

Whildin, J.K., ed. 1860. *Memoranda on the Strength of Materials Used in Engineering Construction.* New York: J.A.H. Hasbrouck.

Wiard, Norman. 1895. *The Solution of the Ordnance Problem.* Philadelphia: J.B. Lippincott.

Williams, Lieut.-Gen. W.F. 1862. *Report of the Officers of the British Army on American Arms.* TNA, WO 33/11/698–740.

Wilson, John Marius. 1848. "Pressure." In *The Rural Cyclopedia; or, a General Dictionary of Agriculture,* vol. 3, pp. 938–939. Edinburgh, UK: A. Fullarton.

Woodbridge, William E. 1856. On the Measurement of the Pressure of Fired Gunpowder in Its Practical Applications. *American Journal of Science and the Arts,* 22:153–159.

FOUR

INFORMATION FLOWS AND FIELD ARMIES

Seymour E. Goodman

In 1860 the U.S. Army had about 16,000 men, mostly organized into regiments made up of infantry companies, cavalry troops, and artillery batteries. These units were posted in permanent or semipermanent installations, with many stationed west of the Mississippi or in coastal fortifications. Administrative and logistical support was simple, modest, and uncontested. A typical field operation at the company or troop level lasted a few days to a few weeks. By the summer of 1864, this had changed. Most notably, William T. Sherman set out to visit Atlanta and beyond with an army group of three armies. A Civil War field army usually had at least two corps; each corps had at least two divisions; each division had at least two brigades; each brigade had at least two regiments, and each regiment had up to 10 companies. With national military command authorities coordinating all field armies, there were up to six levels of command and organization beyond what existed in 1860. And these armies were often moving and in the field for extended periods.

It is useful to define Command, Control, and Communications (C3, pronounced "cee three") as the information flows and attendant technologies necessary to keep field armies effectively operational. The spectrum of C3 ranged from company tactical battlefield commands, with time scales of minutes to hours over distances of tens of yards to a few miles, to the control and management of the movement of armies over time scales of days to months and distances of hundreds of miles. Furthermore, effective logistics required getting these units what was needed where it was needed when it was needed, with corresponding information flows. Dealing with this spectrum required a multifaceted technological transformation involving a hybrid of techniques modified from Napoleonic times along with the then most-modern, widely available information technologies of the evolving industrial era. This chapter tries to describe and assess this transformation, namely, what information flows, via what technological means, were or were not established and turned into reliable operating procedures to enable the necessary scale up and effective conduct of military campaigns.

AN ANALYTICAL FRAMEWORK

By the 1860s, the railroads of the United States were on the way to becoming the "first of the large scale, complex, full-fledged technological systems in the country" (Marx 1998:3). The electric telegraph system, also called the magneto-electric signal telegraph (Beardslee 1863; Plum 1882; Greely 1911a), would be another. But it took the better part of a century before this and later "death of distance" technologies began to coalesce into the complex, interconnected system of systems that has become ubiquitous in today's American military and "information society." The electric telegraph has an important place in this story, but it will be shown to be part of a mix of information technologies used by Civil War field armies, and some more modern terminology will prove useful to discuss them collectively.

The use of modern terminology is not unique to the subject of this study. Indeed, an understanding of medical and sanitation problems during the Civil War is enhanced by an appreciation of post–Civil War scientific discoveries and terminology. More generally, the term *technology*—a key word in the title of this volume—was not much used until after the Civil War, and its use in current, widely inclusive senses did not occur until well into the twentieth century. As Marx (1998:1) states "historians now routinely project the word [technology] back into the relatively remote past." The terms "Command, Control, and Communications" refer to important functions that give rise to many of the information flows essential to the operation of Civil War field armies, and thus it will be helpful to define and project them backward in ways suitable for this analysis.

Terminology

Technology is defined here as the knowhow to conceive of, design, build, operate and maintain useful products and processes. The term also includes the products and processes. Note the importance attached to technology as process. So, for example, a Signal Corps station would have technology in the form of products including telescopes, flags, torches, and codes and ciphers, but it would also embody technology in the form of operating procedures for setting up and dismantling stations, exchanging signals, and making and recording observations.

The term *command and control* (C2, pronounced "cee two") refers to a range of functions that span everything from direct battlefield commands (e.g., "charge" or "hold this position at all costs") to the control of the strategic movement of large units over long distances. The words do not seem to have been used together with any regularity until at least World War I, and to this day there is fuzzy ground between what falls under "command" and what falls under "control" (DOD 1999; McGrath 2006).

During the Civil War, *communications* was commonly used in a military context of "lines of communications," referring to channels or routes between a combat force and its exterior, along which troops, equipment, and supplies flowed toward the combat units, and for the outward flow of the wounded, prisoners, other physical entities, and ultimately for retreat, if necessary. The same protected routes also provided pathways for communications as information flows via couriers or telegraph lines. *Information* takes the form of sounds, images, or text. *Communications* consists of creating or collecting, transmitting, receiving, and processing information. By the mid-twentieth century this third *C* was concatenated into C3.

Intelligence, surveillance, and reconnaissance (ISR) collectively refer to information flows into the field army that are primarily concerned with enemy activity and with topography. In what follows, such functions may be referred to explicitly or be considered as special forms of information flows embedded in C3. By the late twentieth century, after the advent and extensive use of electronic computers, a fourth *C* was added, and all of this was concatenated into C4ISR (DOD 1999:29). Compared with the late twentieth century, the relatively simple circumstances of the Civil War permits a more distinct separation of functions under each of the first three *C*s.

A *field army* is defined here as the unit that included major units of all three combat arms (artillery, cavalry, and infantry) and the necessary military support units of the mid-nineteenth century (notably headquarters, commissary, engineers, medical, ordnance, quartermaster, and signal) that could sustain itself in operations for extended periods. At least one of the large components would be a corps. The field armies fought most of the real *war* of the Civil War. The fewer than 20 designated armies on both sides differed considerably in size and importance.

Two or more armies under a single commander are referred to as an *army group*. The term was not much used to designate an American command until World War II, but it serves as a simple and useful term here. To appreciate why, consider the most complicated field command structure of the war, under

William T. Sherman during 1864–1865. On his trip from Chattanooga to Atlanta he had under his command three armies (Army of the Tennessee, Army of the Ohio, and Army of the Cumberland), and identified himself as the commander of the Military Division of the Mississippi in the Field. After Atlanta fell, he took two armies with him to Savannah (the Army of the Tennessee and the new Army of Georgia) and sent the other two, under George Thomas, after the Confederate Army of Tennessee. These combinations of two or three armies did not have their own names, and Sherman retained overall command under the Headquarters of the Military Division of the Mississippi in the Field, ending up in North Carolina (Sherman 1875). Thus the backward projection of *army group* is convenient for the purposes of this chapter.

A Two-Dimensional Framework

Consider two interwoven dimensions of study presented herein. The first is largely functional, concerned with the purposes of information flows and coupled with the expansion of field units from the company and regimental levels of 1860 to the army groups of 1864. The second is largely technological, considering what technologies were brought to bear to try to solve the functional problems. Particular attention is devoted to the second.

The technologies of both dimensions of this framework evolved rapidly over a short time. For the first, consider tactical (mostly command) and strategic (mostly control) information flows within a field army or with other proximate military units, such as naval units in direct support. For the second, consider external information flows between field armies and more remote senders and receivers—for example, civil and military people and organizations in Richmond or Washington. Both dimensional lines are investigated partly through functions and partly through different time and distance scales.

The second dimension is viewed with an eye on the "two worlds" of the volume title. Which information technologies existed during the Napoleonic Wars, how did they evolve, and how were they used by the time of the Civil War? Which technologies did not exist in any substantial way by 1815, and how did these new information technologies find a functional presence in the Civil War? Which technologies failed to see significant use, and what had to happen before they could or could not find important and extensive uses by World War I? From a military–technological standpoint the American Civil War was indeed astride the two worlds of these earlier and later wars with regard to the information technologies.

Coverage herein is mostly of the Northern armies. This is partly because of surviving records and accounts and partly because the Union had many more resources to work with than did the Confederacy. The differences increased with time as a result of more Northern experience and capacity and Southern attrition and eventual near exhaustion. Among the Union armies, the Army of the Potomac was the most active early user of the then-modern technologies, notably aerial and field telegraphs and balloons, the latter being primarily a platform supporting information flows. Less technologically dramatic, but of great consequence, were the information flows and their attendant technologies in Lee's Army of Northern Virginia in 1862–1865 and Sherman's army groups during 1864–1865.

BATTLEFIELD C3 (TACTICAL)

Civil War combat ranged from frequent skirmishing, sharpshooting, and other small-unit firefights to intense, bloody battles involving entire field armies. Time scales ranged from minutes to a few days. Increased effective ranges of rifle-muskets and artillery permitted firefights on a larger scale than during Napoleon's time, up to a mile or more for these weapons. As the war went on, combat units tended to seek cover and spread out, making C3 more difficult than was the case for the more densely packed and rigidly controlled Napoleonic formations. Different C3 solutions were adopted across this time and distance spectrum. Some battles—notably those involving the capture of cities including Vicksburg, Chattanooga, Atlanta, and Petersburg—were fought over longer distances and time periods and should be considered as strategic rather than tactical operations. Tactical C3 was mostly via technological systems that used sound and sight for transmitting information. Developments in ordnance, including the production of enormous quantities of ammunition, resulted in large, noisy, smoke-filled battles, which made for difficulties hearing or seeing, further compounded by inexperience, carnage, topography, and confusion.

Sound

The most obvious and widely practiced sound form of command and communication was the verbal order, and this was the norm at the company, troop, and battery level. For the regiment, distances and noise background required something else, and couriers came into play, taking oral messages from regimental headquarters to the companies and bringing back situation reports. But distances were often short enough for company and regimental officers to run or ride over to each other for orders and consultations. For infantry regiments, it became common for the colonel to have one company close to headquarters and used partly for these purposes. Messengers would also be used to communicate orally with higher-level and support units.

At brigade and above, tactical field communications usually took the form of mounted couriers or staff officers carrying oral and written messages and staff or commanding officers visiting each other. There are photographs and paintings of groups of couriers and staff officers around senior commanders available to carry out this function. Particularly striking is the panoramic painting by James Walker (2012; also James Ogden, Historian, Chickamauga National Battle Park, pers. comm., 2012) of General Thomas and his staff and couriers in the heat of battle at Chickamauga. The fundamental problem was that effective oral communication, especially against a noisy background, was limited to a few yards, and so the speaker or his agent had to move in order to transmit the sound over longer distances.

This system of couriers and staff officers eventually worked reasonably well for both sides, but it took some time to evolve. John English (1964, chap. 3) describes multiple examples and problems that eventually got sorted out in fairly effective ways as a result of "much bitter experience" by the Confederate armies in Virginia during 1862, including Jackson's Valley Campaign and the Seven Days' Battles. Among the lessons learned was not to let small-unit cavalry commanders select messengers, as they often chose to rid themselves of the least-desirable troopers, who lacked intelligence, courage, or reliability and had some tendency to desert or get lost. By the later years of the war, couriers on both sides tended to be trusted officers. Griffith (2001, chap. 2) has a good discussion of basic tactical C2, including comparisons with Napoleonic armies.

A second category of the use of sound for C3 is via musical instruments, almost always drums or bugles. Bugles worked better in the comparatively quieter and simpler environment of fighting Indians on the frontier than in the heat of large battles. Among the training manuals used during the Civil War are some that explicitly describe musical commands (War Department 1864, plates 81–84; Billings 1982, chap. 9). Drummers and buglers had to learn a large number of military beats and calls, likely more than the average listening soldier could remember. Drums were used more effectively early to control troops in such activities as frontal attacks than later in the war when troops dug in or otherwise sought cover and fought from such positions. Musical instruments were used throughout the war for the control of camp activities such as assembly and reveille.

These "sound systems" satisfy our definition of technology and involve both products and processes. While "low tech" even by the standards of the time, no realistic alternatives existed that could be used as effectively, and none would become available until the creation of the telephone and other forms of electrical voice transmission. Even after the invention of the telephone in the 1870s it took well into the twentieth century for it to become ubiquitous on the battlefield, with whistles and dispatch runners (including Adolph Hitler) extensively used in the trenches of World War I.

Sight

The two most common forms of tactical visual C2 appear often in battlefield paintings and drawings: the prominently displayed commander leading a charge or defense, and a unit following or defending a flag, lining up, or rallying on the national or unit colors. Both forms depend on line of sight and could be severely distance limited by conditions of smoke, topography, or weather. Even so, both were probably at least as effective as shouted orders when the units were fighting or on the move, including marching from one place to another. Everyone on both sides realized the more-than-symbolic importance of these functions, and efforts to interfere with such communications made commanders and color bearers prime targets. Another kind of flag, the signal flag, consisting of a dark-colored square against a white background or vice versa, and the specialized soldiers who used them, became increasingly common in Civil War field armies. Signal flag communications are discussed under "Aerial Telegraphy" after important complementary technologies are presented.

The most advanced visual technologies of the time fall under the scientific heading of optics, using lenses to focus light. Such technologies include evolutionary improvements on optic devices found during Napoleonic times—the telescope and extensible spy glass—and those that did not exist at all or not for practical purposes until after 1815—telescopic sights, binoculars, surveyor instruments, and cameras—and these technologies had both tactical and strategic presences in the field armies. These devices, along with periscopes and heliographs, are all information technologies: they collect data, signals, or images and convert them into useable representations that are transmitted to a receiver. Periscopes and heliographs were not used until later in the nineteenth century, although some Confederate officers used mirrors as crude periscopes in the Petersburg trenches—forerunners of more technologically advanced periscopes used in World War I trenches. Heliographs were used during the post–Civil War Indian Wars, in situations where there were good signaling positions and lots of sun. Although not optical devices, compasses and timepieces are also forms of information technology. Pocket compasses and watches were widely used during the Civil War.

Most telescopes had less than 20× magnification, levels essentially achieved by Galileo in the early seventeenth century. By the Civil War two problems had been solved by better lenses and designs that made it possible for telescopes to be of shorter length and reduced much of the color and other distortions around the edges of images that plagued Napoleonic-era devices. These technical improvements made possible varieties of little telescopes, including telescopic sights used by sharpshooters. The largest concentration of these latter devices was with two regiments of U.S. Sharpshooters in the Army of the Potomac. Sharpshooters became increasingly common on both sides over the course of the war, but the majority did not use such devices. These scopes were of 4× power or less, but, like other line-of-sight optical devices, they had the value of concentrating the observer's attention into a narrow field of vision around the object of interest.

The most important of the optical technologies for Civil War field armies were binoculars. Although binoculars date back to the seventeenth century, the real invention of practical binoculars required the post-1815 technical innovations that, among other things, enabled both tubes to be effectively refocused together. Manufacturing capabilities, in the North, in France, and elsewhere in Europe, were such that many if not most general officers in the field had a pair. They were also commonly used down to captains among artillery, cavalry, engineering, and topographical engineering units. Tactically they were of enormous value over distances of hundreds of yards to a few miles for reconnaissance of the enemy and topography, keeping track of one's own people, and targeting field artillery. They provided depth of vision, a better field of vision, and a better user interface than monocular devices. All of these factors contributed to their rapid acceptance and widespread use on or near the battlefield.

Photography was an optical technology that did not exist during Napoleonic times but was well established during the American Civil War. Movable field units, including equipment for developing and duplicating photographs, accompanied some Northern armies. But there seem to be no significant tactical applications. Photographs of units in combat are rare, although there is a striking photo of what appears to be a Union battery commander using binoculars "under fire" at Petersburg (Miller 1911:16–17). Topographical applications arguably might have been potentially useful, particularly for aerial photographs taken from balloons and quickly developed using the photographic field units. But this was not tried, and it is not clear if the little baskets on the balloons of the time would have made suitable platforms. Photography is revisited in the sections on strategic and external information flows.

Aerial Telegraphy

Both the Confederate and U.S. Congresses officially established Signal Corps for their armies during the war. The Confederate Signal Corps was formally created first, on 19 April 1862, while the U.S. Army Signal Corps was established 3 March 1863. But both the Confederate and Union armies used signal units early in the war, and early practitioners, including Edward Porter Alexander and James E. B. Stuart on the Confederate side, learned from Albert J. Myer in the years before the Civil War. Myer served with the Union army during the war—controversially because his strong advocacy for the newborn branch was not always shared by some superiors—and is now regarded as the father of the U.S. Army Signal Corps.

Field signal units were equipped with visual technologies that included telescopes, colored pyrotechnic devices, binoculars, signal flags, and torches, the latter serving the same functions as flags during the night or conditions of poor visibility. Signaling via flags and torches was often called *aerial telegraphy*. The U.S. Army Signal Corps also initially operated the electrical field telegraph, which is discussed under "Field Telegraphy" later in this chapter. Myer and others developed effective procedures and ciphers and codes for the signalmen to communicate with each other. They also established a training program that produced a steady stream of technically well-qualified soldiers. Many signalmen were very skilled and could send and receive signals at high rates, even given the time it took to physically wave a flag (which was significantly longer than the time to press a telegraph key but still within Civil War tactical time scales). Signal detachments, with field units often called "stations," operated in no less than 12 Federal field armies and military departments, with many departments having a correspondingly named field army, for example, the Department of the Tennessee and the Army of the Tennessee (Brown 1896).

From the perspective of field armies, signal units and personnel had tactical, strategic, and external roles, and a number of examples of units in these roles can be found in all three categories. These examples are briefly taken up under each category in this chapter. Signal officers and men did not normally serve under units smaller than a corps. They were often positioned on high ground, sometimes augmented by the use of towers or trees. From these positions they might have had important tactical communications, intelligence, and reconnaissance roles. Stations could be miles apart, but telescopes permitted clear observation of signal flags, and these communications could be rapid and above the smoke and topographic problems on the battlefield itself. Small signal units, typically with a signal officer and two men, a telescope and a flag, were also used irregularly on the battlefields, sometimes placed in treetops or on roofs. Brown's *Signal Corps in the War of the Rebellion* (1896:135, 228, 420, 434, 455, 469, 477 [Confederate], 479, 495, 531, 579, 655) contains drawings of treetop signal stations in most of the Federal military departments or field armies. The Atlanta Cyclorama depicts a combat-explicit example: in the middle of the raging conflict stands a tree with a ladder and makeshift steps nailed much of the way up the trunk, from which hangs a large, shredded, red-on-white signal flag; the caption reads "the tree with the flag affixed is a Federal signal station" (Kurtz 1954:14).

For an important tactical example, Brown (1896:330–333) describes the positioning of signal stations on mountains near Antietam, including one on Elk Mountain, where

> a full view of the enemy's lines was here obtained. Every change in
> position, the shifting of batteries, and the evolutions of cavalrymen,
> were distinctly witnessed and reported. At the same time our own
> signal stations were visible and, in consequence, the information
> here secured could be delivered in all parts of the Army.

Brown then goes on to describe some of the actions of these signal stations during the Battle of Antietam. One can imagine that such a well-placed network would provide the Army of the Potomac with a huge communications, intelligence, surveillance, and reconnaissance advantage. As testimony to that effect, Greely (1911b:320–321) has photographs of the Elk Mountain station and cites a Confederate eyewitness correspondent from Richmond:

> Their signal stations on the Blue Ridge commanded a view of every
> movement. We could not make a maneuver in front or rear that
> was not instantly revealed by keen lookouts; and as soon as the
> intelligence could be communicated to their batteries below, shot
> and shell were launched against the moving columns. It was this
> information, conveyed by the little flags upon the mountain-top,
> that no doubt enabled the enemy to concentrate his force against
> our weakest points and counteract the effect of whatever similar
> movements may have been attempted by us.

The Antietam case notwithstanding, it is doubtful if such advantages were frequently realized or most effectively exploited. It is one thing to have a network as described among signal stations, but that is only part of the transmission picture. Valuable, often time-sensitive, information had to go from the closest signal stations to battlefield commanders who were in a position to act on that information and convey commands that would convert the intelligence into battlefield results. It might also be helpful if there were procedures whereby senior ground commanders could convey pressing requirements to the signal stations. It appears that signal units were not systematically built into the tactical operations of the field armies to the extent that their activities became part of normal procedures widely appreciated or utilized by commanders at any of the command levels in most armies. It also may have been the case that, although short by some measures, the times for these communications, intelligence, and reconnaissance processes were too long for effective battlefield integration.

The great majority of tactical examples come from former Union signalmen like Brown and Greely. Although they tried to acquire information from former Confederate signalmen, with the exception of a few sources such as early efforts described in Porter Alexander's memoirs, there is little information on the Confederate Signal Corps, or at least little has survived (Alexander 1989; Cummins 1888; Brown 1896, chap. 9). There is no evidence of a training program producing large numbers of capable signalmen, as there was in the North, and Confederate tactical use seems to have been far less extensive than that of the Union forces.

Balloons

Both Northern and Southern armies used balloons as platforms for observation, reconnaissance, and information transmission. There were potentially great observational advantages to be sought several hundred or more feet above battlefields on platforms that were more portable than mountaintops. These platforms could be exploited using the much-improved telescopes and binoculars. Communications with the ground were through written messages dropped or slid along ropes or via field telegraph for air-to-ground communications. Lowe (1911:380) describes circumstances whereby he got an electric telegraph aloft on the balloon *Intrepid* near Fair Oaks on 31 May 1862 and kept "the wires hot with information." Haydon (1941:329–330; see also Zeller 2005:67–68) describes other times telegraph transmission was used with balloons. Photographs had been taken from balloons but not during their field deployment during the Civil War. Aerial photography had been suggested to Lowe, but he apparently never tried. Since observation balloons were also observable by the enemy, they had secondary value in that their presence might influence the behavior of that enemy by forcing greater efforts at concealment or the use of longer routes to avoid observation, thereby resulting in some intelligence against counterintelligence gaming (see Macaulay, chap. 8, this volume).

Balloon use was mostly in eastern and northern Virginia, most notably during the Peninsular Campaign and around Fredericksburg (Lowe 1911). In spite of the advocacy of Thaddeus S. C. Lowe and others, balloons never amounted to much and as such do not receive further attention here except to note some technical factors and their place "astride two worlds." Civil War balloons were vulnerable to weather conditions, lacked means of controlling their movement aloft, and had to be strongly tethered, at least partly to keep them from spinning around. They also required a great deal of ground support to move them and generate the hydrogen-based gas to keep them aloft. Confederate balloons had to be filled in Richmond. Had they been more effective and had there been more inclination to use them, the Union armies would have had a near monopoly. Given comparative manufacturing capabilities and skilled personnel, the Union armies might have maintained a balloon unit with some corps, and certainly with the major field armies, whereas the Confederate armies would have been hard pressed to do anything beyond their very limited efforts early in the war. In this regard it is hard to resist retelling the most colorful Confederate balloon story. Lowe (1911:382; see also Bradlee 1925:298–299) quotes Confederate Gen. James Longstreet after the war:

> The Federals had been using balloons in examining our positions, and we watched with envious eyes their beautiful observations as they floated high up in the air, well out of the range of our guns. While we were longing for the balloons that poverty denied us, a genius arose for the occasion and suggested that we send out and gather silk dresses in the Confederacy and make a balloon. It was done, and we soon had a great patchwork ship of many varied hues which was ready for use in the Seven Days' campaign. We had no gas except in Richmond, and it was the custom to inflate the balloon there, tie it securely to an engine, and run it down the York River Railroad to any point at which we desired to send it up. One day it was on a steamer down on the James River, when the tide went out and left the vessel and balloon high and dry on a bar. The Federals gathered it in, and with it the last silk dress in the Confederacy.

Manned balloon flights were established in France in the 1780s. The idea of portable platforms at great heights quickly stirred imaginations on their possible military applications, including observation, troop transport, long-distance signaling, the dropping of propaganda leaflets, and bombardment. There were experiments and speculations in France and other European countries. Flights of a thousand miles and heights of 20,000 feet were achieved. But balloons ultimately did not have military roles of great consequence (Haydon 1941, chap. 1). During the Civil War only observation missions were seriously pursued. Balloons did not achieve the potential envisioned by Lowe and others until World War I, when a large number of observation balloons were used along the trench systems of the Western Front. In both World Wars balloons were used to discourage low flying aircraft, and as well-controlled "blimps" for various purposes, notably long over-ocean patrols.

Field Telegraphy

The U.S. Army Signal Corps developed and deployed a movable field telegraph, as "some appliance was necessary to place adjacent bodies of troops in communication when the topography of the country would not admit of aerial signals" (Brown 1896:171). The equipment necessary for a working system, including batteries, poles, and reels of insulated wire for around five miles, was packed up into a battery wagon and other wagons to move the entire kit and team. Well-trained teams could deploy a working system within several hours.

English (1964:103–114) recounts attempts to use the field telegraph in a strong tactical sense, that is, in near real time during a battle, by the Army of the Potomac during the battles of Fredericksburg and Chancellorsville. English ventures that at Fredericksburg,

> from a technical viewpoint the assaulting Union army was probably the most modern that the world had ever seen. It was capable of communicating by semaphore signals and of conducting aerial reconnaissance. It was also the first to employ the magnetic telegraph as a medium of communication on the battlefield.

Although portable by Civil War standards, the field telegraph still required at least several hours to move, was dependent on miles of wires, and thus largely limited in a tactical setting to connecting high-level headquarters to other headquarters (and perhaps limiting the mobility of some commanding officers by the standards of the time) or to important, stationary positions like bridgeheads. The difficulties of maintaining wires against malicious or accidental disruption were considerable.

English (1964) tries to make the case that inexperience and other factors resulted in misuse, including overuse, of these technologies during both battles (Fredericksburg and Chancellorsville) by the Union generals and that Lee and his subordinates used less-sophisticated technologies, mostly their well-developed courier system, more effectively. These were undoubtedly secondary factors in the outcomes of both battles. It is unlikely that these technologies, however effectively used and appreciated, could make up for the more fundamental differences between the plans and their execution by the opposing commanders. It would take until 1864, in particular during the relatively stable and extended conflict around Petersburg, for the field telegraph to be used more extensively and effectively.

There was no established field telegraph in the Confederate army. An attempt was made by Confederate Postmaster John Reagan to provide Lee with a prototype, but "it was found that the broken character of the country, together with the extent of the forests in which the Army of Northern Virginia operated precluded the successful operation of the apparatus" (Bradlee 1925: 298).

The field telegraph is notable here partly because it is neither a sight nor a sound technology but one of the few applications of electricity and magnetism on the battlefield in the Civil War. British physicist James Clerk Maxwell developed his unifying theory for electricity and magnetism during the years of the Civil War, but it was of no consequence to the war. Although not usually thought of in this sense, the magnetic pocket compass was the most widely used such technology. Another battlefield use of electricity was as a fast wire fuse to set off explosives.

CONTROL OF FIELD ARMIES
(STRATEGIC C3IR)

Strategic C3IR pertains to the control and management of the movement of armies and their supplies to bring about battle under favorable conditions or to cause the enemy to retreat or to otherwise diminish the enemy's ability to wage war. It involves communications and information flows over distances and times greater than those of the typical Civil War battle. Distances for movements and operations could extend up to hundreds of miles. Times could extend up to months. Strategic C3IR covers most of what armies did in the field when not in battle or sitting in camp. During such movements, relatively minor forms of combat would take place, for example, clashes between cavalry units or skirmishes during reconnaissance operations.

Four particular categories of strategic operations fall within the general control of field armies. The first covers the taking of cities over an extended period, that is, sieges, notably Vicksburg, Chattanooga, Atlanta, and Petersburg. These campaigns would often include relatively short periods of intense, large-unit combat, for example, the Crater at Petersburg.

The second category includes movements where a large unit of a field army is separated from the main body by a considerable distance. This would include operations in the Shenandoah by corps under Jackson and Early at different times during the war (in 1862, Jackson's command had a weak field army designation as the Army of the Valley) and by the Army of the Potomac (notably under Sheridan); Jackson at Harpers Ferry during the Army of Northern Virginia's Maryland campaign in 1862; and the separation of armies within army groups. Examples of the last include Sherman's two army groups under the Military Division of the Mississippi in the Field and the two separated armies under Grant near Richmond (the Army of the Potomac under Meade and Butler's Army of the James). A sub-category would be a large raid, when typically a cavalry brigade or division would be cut loose for reconnaissance in force or to accomplish another mission that would hurt the enemy or directly benefit the raider's army. Stuart's rides around McClellan in 1862 and before Gettysburg in 1863 are notable examples.

The third category of strategic operations involves long lines of communications and information flows that had to be operated and protected by troops from the field armies. An important case is that of Sherman's march from Chattanooga to Atlanta in mid-1864. Another is Grant's solution to logistics problems at Petersburg by building a major port at City Point and extensive lines of communication from there to the Army of the Potomac and (to a lesser extent) to the Army of the James.

The fourth category includes ship-to-shore communications when naval and other ships were in close support of a field army. As would be the case in World War II, the U.S. Army owned or leased more ships than the U.S. Navy. These included transports and small craft.

Key information flows are concerned with knowing where friendly and enemy forces are, where the friendly forces are supposed to be going, how to get them there under time constraints, and how to get them there properly supplied and in a condition to successfully engage the enemy. Since the Civil War field army was a large, complex physical entity composed of a considerable number and variety of units with many different characteristics that determined their value and affected their mobility, these were enormous problems.

These problems were compounded by the physical conditions under which armies often operated. Throughout much of the Confederacy, rough topography, dense forests, swamps, rivers, and other obstacles abounded, making movements difficult and dangerous. Road networks were sparse, the roads themselves often crude and poorly mapped; they suffered terribly in bad weather or after the passage of large units with thousands of men and animals, making it more difficult for those who followed. An important information flow to deal with this was the creation, distribution, and effective use of maps. The success of Stonewall Jackson's Shenandoah Valley campaign of 1862, for instance, owed a good deal to the expert maps of his topographer, Jed Hotchkiss. Jackson's missteps in the subsequent Peninsular Campaign may be partly attributed to the absence of Hotchkiss and his maps (Sears 1999:7; see also McDonald 1973).

Many maps were made before the start of the war for a variety of reasons, such as land and property surveys and the U.S. Coast Survey. Some commanders, notably Sherman, Thomas, and Farragut, were particularly good at acquiring and using such maps. The quality of these maps was not generally better than those that existed during Napoleon's time. During the war, maps were made by people called cartographers, topographical engineers, or just "mapmakers." Some became highly skilled professionals. Topographical engineers were among the most important and frequently consulted officers on an army commander's staff. The maps they produced ranged from crude sketches to cartographic works of art. First and foremost, these maps had to show basic relative layout of the land, roads, and human-made and natural features that would serve to orient and mark progress so that field commanders could move their troops without getting lost or in each other's way. Civil War maps were often weak in showing accurate scale or elevations. Their use had to be augmented with compasses and binoculars, sometimes with scouting parties or local guides. Confederate armies could usually rely on local residents or soldiers who had lived in the area; the Federals often called on slaves or freedmen. But the effective control of armies on the move was sometimes at risk without these people.

Once good maps had been created, they had to be gotten to officers who would use them in conjunction with the army commander's plan to move and place people and equipment. That meant that the maps had to be reproduced and distributed down to the necessary command levels. Both sides used several methods of map reproduction. Techniques included simple copying by draftsmen, tracing methods, chemical treatment and exposure to the sun as a form of "solar printing," various forms of lithography, and photographic reproduction (McElfresh 1999, chap. 12). As always, the Confederates were at a disadvantage with regard to any technique that required advanced equipment or materials.

Written orders or movement plans would also be duplicated and distributed. Text was more easily duplicated than were maps. As Joan Boudreau describes (2012), portable printing presses were an established technology and sometimes available at field army headquarters. They were also used to support other information flows through the production of requisition forms, unit newsletters, and parole documents. Both sides had soldiers who had been printing tradesmen before the war. Distribution of the maps and other documents was not always easy. They had to go to the right people in timely ways, which meant knowing where these people were. This was done person to person, usually via courier or staff officer or at meetings between commanders and their subordinates. Obviously maps could not be transmitted by either aerial or electric telegraph.

The aerial telegraph was used more extensively for strategic purposes in some armies than others. The most extensive was in the Army of the Potomac, which also was close enough to Washington for Signal Corps stations to sometimes serve as communications links all the way to the capital. In the West, the Army of the Cumberland under George Thomas was the most extensive user of aerial telegraphy. The 100,000 men with Sherman from Chattanooga to Atlanta included only 200 signal troops, almost all with Thomas (Sherman 1875:16–22).

The electric telegraph had more prominence strategically than tactically, although less than it did for external information flows (discussed later in chapter). The time and distance advantages of the electric telegraph were so evident that many field army commanders and subordinate functionaries used the telegraph as much as possible. This was facilitated by the existence of a substantial long-distance network and supporting infrastructure for the united country at the beginning of the war, and the armies often found physical facilities in place or readily repairable as they moved around their theaters. The existing infrastructure was sufficiently dense and reliable that the field army commanders and their subordinates were able to count on the availability of the people and organizations on the other ends of their communications. It was not uncommon for army commanders to construct short links from their headquarters to main telegraph lines. Both Northern and Southern field army commanders had troubles with "ownership" of the telegraph lines and operators within their geographical domains, which could handicap effective telegraph use. Competing for ownership and priorities were other specialized military organizations or civilian companies (Miller 2012).

The strategic value of the electrical telegraph networks was demonstrated early by Confederate use to bring troops some distance to the First Manassas battlefield in time to contribute significantly to that victory. Use of the telegraph in strategic operations was facilitated by relatively stable geographic situations. "Stable" does not mean static but refers to situations where the armies could count on using telegraph routes for extended periods of time. An important case is Sherman's use of the telegraph and railroad corridor between Chattanooga and Atlanta. Although contested by Confederate forces and requiring the commitment of thousands of troops to protect these lines of communications, they were successfully maintained and proved critical to that campaign.

The telegraph was used extensively in other field armies as well. Logistics and the communications for logistics had been well developed for the Army of the Potomac over most of the war. It helped that this was the largest, most visible of the Union's armies and that it was located relatively close to some of the North's largest port cities and manufacturing centers. But at Petersburg something new had to be established to support a position south of Richmond. The result was a large port at City Point, Virginia. Lines of communication to the Army of the Potomac included railroad and telegraph facilities, signal stations, and roads. Large numbers of troops had to be detailed to build, operate, and protect these lines of communications, something permitted by the relatively stable nature of most of the Petersburg campaign. Throughout its existence the Army of Northern Virginia used its interior position within Virginia to maintain telegraphic contact with Richmond and other parts of the state.

Arguably much of the value of the telegraph for strategic C3I, and especially for the control C, related to the location and movement of troops and equipment, as was the case for the Confederates for the First Manassas. When available in a stable form, it could be used for these purposes between the main body of a field army and a large detached unit. Timely information on the locations of one's own forces, intelligence on the location of enemy forces, and instructions for movements could be exchanged. Not that this always produced positive results, but even failures help illustrate the value of this form of information technology.

Two cases illustrate the second category above, where large units of armies were separated from the main bodies by 10 miles or more. In both, the long-distance telegraph might have come close to putting the army commander in near-real-time tactical and strategic C2 over the separated forces. General John Bell Hood should have been in essentially continuous telegraph contact with his troops in the Confederate Army of Tennessee protecting the railroad at Jonesboro during the campaign for Atlanta. He arguably had better C3 with his forces there (including a direct rail link) than Sherman had with the army he was using to make a wide sweep to cut off that last rail line into Atlanta. Nevertheless, Hood did not appreciate what was happening until too late, and the failure forced him to abandon Atlanta. The second case occurred later in 1864 when Hood took his army north to Tennessee against those of Schofield and Thomas. The two Northern generals should have had excellent telegraph communications from Franklin to Nashville, which could have given Thomas timely C2 influence over the impending battle at Franklin. But just before that battle, Schofield's telegraph operator deserted (why no backup was immediately available is not explained).

Neither General Thomas, Schofield, nor any of their staff officers were permitted to know the telegraph code and the result was that from eight to forty-eight hours were occupied in sending a dispatch that should have gone through in a few minutes. And this in the most critical days of a critical campaign. (Bradlee 1925:295)

Nevertheless Schofield and his troops did well at Franklin, thereafter joining Thomas in Nashville, where they effectively put Hood's army out of action.

The overall northern telegraph infrastructure and its supporting industry permitted it to maintain and extend telegraphic lines over thousands of miles, including the territory that the Confederacy lost during the war. The telegraphic capabilities of the South were steadily reduced, with few new lines and the deterioration and loss of much of what they inherited from the Union when the war began. But southern telegraphic connections were never reduced to zero.

The most common, and ultimately effective, information flows for the control of field armies on the move were via couriers, staff rides, and commanders visiting with their subordinates (e.g., Jones 2005). This worked best when supported by good maps, binoculars, and compasses. This was always the case for Confederate field armies, as they had little or no opportunities or resources to experiment with new technologies. For the most part, this process was fairly well established by the end of 1862 and worked reasonably well (English 1964). But there were notable exceptions, as when Stuart and his cavalry disappeared from Lee's ken shortly before Gettysburg. There were other cases, for both North and South, where large bodies of detached cavalry more or less disappeared from their parent field army commanders. It was difficult for a mobile force deep in enemy country to maintain effective contact with its parent and risky for couriers, and the information they were carrying, trying to work between the two.

A particular challenge, given the distances involved and issues of lines of communications, was Sherman's march from Atlanta to Savannah and from there to North Carolina via South Carolina. After Sherman took Atlanta, Jefferson Davis spoke in Macon, predicting that wherever Sherman might go the vulnerabilities of his lines of communication were such that he would be defeated as Napoleon was in Russia in 1812. Grant's reported response was "Who would furnish the snow?" (Scaife 2004:40). By this time Sherman had become a thorough planner and implementer. He "kept couriers going to and fro" (Sherman 1875:272) to coordinate his two wings and keep them moving on time and along the planned routes. With due credit for doing a masterful job, what else could he have done? The fundamental limitations and deficiencies of all of the more modern technologies of the time—electric and aerial telegraphs, balloons, and railroads—left none suitable for the requirements of that kind of movement. Homing pigeons ("flying telegraph") might have worked in one direction but were not used by the U.S. Army until 1878 (Spears 1947:535).

The two most likely uses of photography for strategic C3IR would have been for taking photos of one's own or enemy positions and physical features and for duplicating materials, for example, maps and orders, for distribution within a field army command. By 1864, the quality of outdoor photography was impressive (Miller 1911; Barnard 1977; Zeller 2005). But few photographs seem to have been taken leading up to a battle, although many were taken afterward and had little to do with a field army's strategic C3IR needs. Given the technical limitations of the time, the value of photography to strategic IR operations would have to wait for twentieth-century photographic equipment and observation platforms.

But photographing orders and maps for wide distribution among unit commanders of a field army would seem to be a fast and uniformly exact way to accomplish this important function. Some photographic reproduction of maps was done in the North. A photograph of Alexander Gardner's "secretive" facility for "map photographing for the army in the field" is shown in King (1911:23). The drawbacks were that equipment was bulky, sun conditions had to be just right to make reproductions, and photographic copies distorted at the edges. This last drawback made it difficult to connect maps of adjacent sections because the roads and streams were distorted where they were supposed to meet. Finally, the use of the photographic copies in sunlight caused them to fade (McElfresh 1999:69).

But this does not fully explain why text documents, such as orders, were not photographically copied and distributed. Apparently less technologically advanced methods sufficed. Once again, these uses of modern technologies apply almost exclusively to the U.S. Army. The supply of photographic equipment and chemicals was problematic for the Confederates. As the war dragged on there were fewer and fewer southern-originated photographs of any kind.

Throughout the war, Federal field armies engaged in operations along coasts and large rivers and were in communication with ships. The most prominent of these operations included the "river wars" in the West, particularly along the Tennessee, Mississippi, Cumberland, and Red Rivers; the 1862 Peninsula Campaign; and various operations along the 3,500 miles of Confederate coast. Ships also supported such large-scale supply and troop transportation operations in 1864–1865 as Sherman's advance from Savannah through the Carolinas and Grant's operations in Virginia.

Navy and army ships worked directly with the field armies, including providing direct fire support, transporting and landing troops and equipment, and providing a means for communications for dispatches, letters, and couriers to distant places. Much of this cooperation was worked out on a case-by-case basis and went particularly well when the army and naval commanders established good direct working relations.

This interaction required forms of ship-to-shore communications on an unprecedented scale and usually was done in two ways. The first was for people to shuttle between ship and shore. The second was via aerial telegraphy, using signal flags by day in good visibility and torches and lanterns by night and conditions of poor visibility. Both the U.S. Army and U.S. Navy used flag and torch technologies of their own during the war, and ship-to-shore and ship-to-ship communications were widespread in navies throughout the world. The army and navy systems were not identical. The more compact army system used rapid motions of a single flag or torch, whereas most naval systems used multiple flags or colored lanterns in relatively fixed displays. Often this difference was handled by having small Army Signal Corps teams work on the ships, communicating with their own kind on shore. One of the most interesting operations involving ship-to-shore signaling was naval gunfire to support the Army of the Potomac around Malvern Hill and Harrison's Landing against moving Confederate land forces in mid-1862. This gunfire was directed in real time via aerial telegraphy using teams on ships and shore (Brown 1896:318–321).

Technology for underwater electric telegraph cables had been developed before the war and was used across rivers and along coasts, but this possibility for ship-to-shore communications was not used to any extent, and there seems to have been no pressing need for it. For the most part ship-to-shore distances and time scales were just about right for the flags and torches and optical devices of the Civil War period. It helped that these operations and communications were essentially uncontested by the Confederates.

Finally, the industrial era necessitated expanded bureaucratization of the military with accompanying official paper flows. Among these were after-action reports, reports on expenditures of ammunition by artillery batteries, order forms and receipts for supplies.

EXTERNAL INFORMATION FLOWS

Information flowed not only within and between field armies but also externally. Two categories of long-distance external information flows may be identified. The first is official, including direct communications between the field armies and distant military and civilian authorities, with regional military district headquarters, state capitals, and logistical depots. The *Official Records of the Union and Confederate Armies* (in *The War of the Rebellion*, War Department 1880–1901) are filled with such communications. The second is unofficial, comprising a wide variety of information flows between officers and men of all ranks and civilians travelling with the field armies (e.g., photographers, newspaper reporters, and sutlers) on the one hand and with people throughout northern and southern civilian populations on the other.

Communications in both categories were carried via three primary networks on both sides. These were the long-distance telegraph, couriers, and the national postal systems. Although occasionally used as links in some of the information flows in the official category, the aerial telegraph had a small role in such communications. Early versions of all three long-distance networks existed in France from the 1790s and were extended during the Napoleonic wars (Hubert 2012). An aerial semaphore telegraph was invented by Claude Chappe in 1792 and operated through the Crimean War. By the 1840s, within the current borders of France, the network included over 500 relay stations and about 3,100 miles of line-of-sight links between them. During the Napoleonic wars the system was extended into neighboring countries and imitated by other European powers. This system was more extensive and sophisticated than the one Albert Myer would create for the Signal Corps, but the latter was better suited to the field needs of the armies of the 1860s.

The French postal service was well developed and used for state and military communications. It included an express system more sophisticated than the later American Pony Express. Napoleon also had a formal *officiers d'ordonnance* and military couriers who operated at tactical, strategic, and external distances, including from Paris to Moscow in 1812, a trip taking two to three weeks. But Napoleon had neither the railroad nor the electric telegraph that were to transform all three networks by the 1860s.

The railroad networks of the North and South supported all three information networks. The long-distance telegraph lines often followed the railroad rights-of-way, with telegraph offices located in or near railway stations. This proximity of railroads capable of carrying heavy materials and work crews also facilitated the construction and repair of telegraph lines. The same physical routes for lines of communication that supported the paper trails for logistics were also routes for the transport of supplies. Long-distance couriers were usually staff officers carrying written and oral communications and travels to Washington, Richmond, or elsewhere were much facilitated by riding the rails. The huge increase in military command levels and general bureaucracy and an unprecedentedly literate population produced large volumes of official and personal letters and packages. The postal systems of both the United States and the Confederate States came to carry much of this mail between the nearest railhead for the troops in the field armies and their correspondents. The river and coastal water transportation networks supported these information networks as well.

It is at this level of information flow that the electric telegraph became very prominent and important for the field armies. Nothing else could move text information nearly as quickly and sometimes close to door-to-door (but senders and receivers of any importance never did the keystrokes themselves, so there were operators and messengers with pieces of paper at both ends). Most of the communication traffic in and out of the field armies was official. It included traffic from the army commanders to the presidents and war departments.

As with other forms of information technology that required substantial equipment, materials, and expertise, during the course of the war the North got stronger and the South weaker in their abilities to use, maintain, and extend the telegraph. The North constructed thousands of miles of new lines, some of which kept up with the movements of Union field forces. Northern capacity was such that it could extend the telegraph into the far west, making the Pony Express obsolete before it had much of a life. The South had difficulties repairing damaged, much less building new, lines and training personnel to sustain and operate them. Sherman destroyed some 500 miles of telegraph in Georgia. While he was doing that, Confederate lines were abuzz with "Where is Sherman?" traffic (Bradlee 1925:305–306). Most Southern armies east of the Mississippi, however, managed to keep minimal telegraph contact with Richmond as long as they effectively operated in the field. Both aerial and electrical telegraph traffic were fairly easy to intercept, and each side tried to read the other's messages. The North was better at making and breaking ciphers. Efforts were made at spoofing and direct attacks. For examples, see Bradlee (1925, chap. 4) and Fishel (1996:347–348).

The postal system, with a history going back to colonial North America, was the first national information network of great consequence. As the 11 Confederate States seceded, Confederate Postmaster General John Reagan wisely and largely seamlessly incorporated everything he inherited from the Union into the Confederate system. The transition included keeping many post offices and routes and honoring U.S. stamps (Rifkind 2010). During the Civil War, the U.S. Post Office expanded and even introduced the first home delivery. The Confederate Post Office managed to function to the end, subject to the stresses of diminishing territory and every other resource needed to transport and deliver mail.

Although I was not able to locate detailed national level data on literacy rates among the troops and mail volumes, the large number of surviving letters indicates that both were substantial. This was the first major war where a large fraction of the officers and men were at least functionally literate. The official paperwork for running the field armies was enormous, and much of it found its way into the mails. The volume of images, notably maps, drawings, newspapers, and photographs, were such that the mails had to be a primary carrier. The Civil War was the first war where men of all ranks engaged in a voluminous correspondence, most importantly with family. Newspapers, food, socks and other physical items were also sent in volume to the field armies. These items were also carried to and from the armies by soldiers on furlough, a distributed irregular postal service. I was unable to find data on delivery problems, including what fraction of correspondence may have been lost in transmission, particularly in the South.

Within the field army, the process of writing and receiving letters was fairly well established. The soldier would acquire writing materials and stamps from home or a sutler. The finished letter would be turned over to the regimental sergeant major, who would deliver it to a brigade collecting point; from there it would go to the nearest railhead. Letters to the soldiers would arrive at the nearest railhead, then be removed to mail tents and sorted by civilians or men from the brigades, and the process reversed until they usually got to the intended recipients. Mail traffic to and from the Northern field armies took about three days. King (1911:33–35) contains photos of mail tents and corps-level mail wagons.

One might imagine the importance of mail call to the officers and men. Receipt of letters was the primary direct connection between soldiers and the external civilian populations. It affected troop morale, concerns about family conditions at home, and desertions. All three were particular problems for the Confederate armies late in the war, as soldiers who were the only (or only surviving) adult male member of poor families learned of the increasingly desperate hardships at home. In many letters the men and their relatives write of the importance of receiving mail. There appears to be, however, surprisingly little on mail call in some of the best known books on the lives of the average soldier, Billings (1982, chap. 11) being an exception. But one sentence in the last paragraph of Sherman's memoir (1875:408), in the final chapter entitled "Military Lessons of the War," says it: "Lastly, mail facilities should be kept up with an army if possible, that officers and men may receive and send letters to their friends, thus maintaining the home influence of infinite assistance to discipline."

The American Civil War was chronologically almost equidistant between the end of the Napoleonic Wars and the start of World War I. It is also between the two in terms of new and widely available information technologies. Most notably the transition from 1815 to 1861 included binoculars, photography, and electric telegraph; and that from 1865 to 1914 included telephone and radio.

Much of what has been described for tactical C3 had Napoleonic precedents. Both sides, but especially Confederate forces, had little technology-enabled alternatives and a great deal of "do or die" pressure in the form of rapidly scaled-up warfare forcing them to adapt and implement processes from Napoleonic times. Arguably none of the post-1815 technologies played a major role at the tactical level in the Civil War with the exception of binoculars. Balloons and the field telegraph were developed, however, to the point where the Army of the Potomac tried them in battle. Adoption of the field and aerial telegraphs at the tactical level suffered from the need to have highly trained and skilled people at both the sending and receiving ends. These, and other technical constraints and handicaps, severely limited the number of such units, and few worked their way down to the division or lower levels. They made for new, strange, and difficult "user interfaces" for hard-pressed battlefield commanders.

A real revolution in tactical C3 would have to await electromagnetic-based technologies that were not even remotely available during the Civil War, namely voice transmission and radio, to overcome these impediments. The telephone was invented in 1876, at least as decided by the U.S. Patent Office. The first military application known of was in 1877, by Royal Engineers on the Northwest Frontier of India. Although both telephony and radio would be created and used before the end of the nineteenth century, it took until World War II for these and other electromagnetic-based technologies to attain widespread and critical importance on the battlefield. Neither telegraph nor telephone achieved ubiquitous tactical use in the trenches of World War I because the wires they still required were much at risk on the battlefield, although there were battery-supported field telephones later in that war. So Adolph Hitler and other dispatch runners still had plenty to do. Although radio was well established by 1914, two-way voice communication over radio was far less so; radio on the *Titanic*, for instance, used telegraphy, not telephony.

As discussed at length the changes were greater for strategic Civil War C3 and for external communications. Still, some of the most effective means were significant modifications of Napoleonic-era technology as processes rather than the use of the most then-modern technology products, the electric telegraph being the important exception. During both the Civil War and World War I it can be argued that the "death of distance" technologies (telegraph, telephone, and radio) had their most important roles outside the internal operations of the field armies. The various tactical and strategic C3 "solutions" employed during the Civil War varied considerably among the field armies and their commanders. The latter had considerable discretion and limitations over uses within the armies. External information flows were more institutionalized and uniform.

The advances in C3 that took place during the Civil War were of enormous importance. Both sides started near zero with C3 or logistics above the regimental level but within two years were effectively operating multiple field armies under contested conditions over vast distances for two remaining years of a very serious war. By comparison the Great Powers entered World War I in 1914 with huge armies and much forethought by general staffs, but within a year they made a bigger and deadlier mess of things than did the American Civil War generals.

References

Alexander, Edward Porter. 1989. *Fighting for the Confederacy,* ed. Gary W. Gallagher. Chapel Hill: University of North Carolina Press.

Barnard, George N. 1977. *Photographic Views of Sherman's Campaign (1864–65).* New York: Dover.

Beardslee (Magneto-Electric Company). 1863. *Directions for the Practical Working of Bearslee's Magneto-Electric Signal Telegraph.* New York: Beardslee. Reprinted (n.d.) Augusta, Ga.: U.S. Army Signal Corps.

Billings, John D. 1982 [1887]. *Hard Tack and Coffee; or, The Unwritten Story of Army Life.* Reprinted Alexandria, Va.: Time-Life Books Collector's Library of the Civil War, 1982. [First published Boston: George M. Smith, 1887.]

Boudreau, Joan. 2012. The Portable Press and Field Printing during the American Civil War. *Printing History,* new series, 12:3–26.

Bradlee, Francis B.C. 1925. *Blockade Running during the Civil War and the Effect on Land and Water Transportation on the Confederacy.* Salem, Mass.: Essex Institute. Reprinted Salem, Mass.: Literary Licensing, 2011.

Brown, J. Willard. 1896. *The Signal Corps in the War of the Rebellion.* Boston: U.S. Veteran Signal Corps Association. Reprinted Baltimore: Butternut and Blue, 1996.

Cummins, Edmund H. 1888. The Signal Corps of the Confederate States Army. *Southern Historical Society Papers,* 16:93–107.

DOD (Department of Defense C4I Plans and Programs, Committee to Review). 1999. *Realizing the Potential of C4I: Fundamental Challenges.* Washington, D.C.: National Research Council, National Academies.

English, John Alan. 1964. Confederate Field Communications. Master's thesis, Duke University, Durham, N.C.

Fishel, Edwin C. 1996. *The Secret War for the Union: The Untold Story of Military Intelligence in the Civil War*. Boston: Houghton Mifflin.

Greely, A.W. 1911a. Telegraphing for the Armies. In Miller 1911, pp. 341–368.

———. 1911b. The Signal Service. In Miller 1911, pp. 13, 305–340.

Griffith, Paddy. 2001. *Battle Tactics of the Civil War*. New Haven, Conn.: Yale Nota Bene.

Haydon, F.S. 1941. *Aeronautics in the Union and Confederate Armies*, vol. 1. Baltimore: Johns Hopkins University Press. [Vol. 2 apparently never published.]

Hubert, Tanguy. 2012. *The Chappe Telegraph, and Communications in Europe in the First Half of the 19th Century*. Atlanta: Sam Nunn School of International Affairs, Georgia Institute of Technology.

Jones, R. Steven. 2000. *The Right Hand of Command: Use and Disuse of Personal Staffs in the American Civil War*. Mechanicsburg, Pa.: Stackpole.

King, Charles. 1911. Military Information and Supply. In Miller 1911, pp. 18–36.

Kurtz, Wilbur G. 1954. *Atlanta Cyclorama, The Battle of Atlanta*. Atlanta: City of Atlanta.

Lowe, Thaddeus S.C. 1911. Balloons with the Army of the Potomac. In Miller 1911, pp. 369–382.

Marx, Leo. 1998. "The Invention of 'Technology.'" In *Major Problems in the History of American Technology: Documents and Essays*, ed. Merritt Roe Smith and Gregory Clancy, pp. 2–6. Boston: Houghton Mifflin.

McDonald, Archie P., ed. 1973. *Make Me a Map of the Valley: The Civil War Journal of Stonewall Jackson's Topographer*. Dallas, Tex.: Southern Methodist University Press.

McElfresh, Earl B. 1999. *Maps and Mapmakers of the Civil War*. New York: Harry N. Abrams.

McGrath, John, J. 2006. *Crossing the Line of Departure: Battle Command on the Move, A Historical Perspective*. Fort Leavenworth, Kans.: Combat Studies Institute.

Miller, Francis Trevelyan, ed. 1911. *The Photographic History of the Civil War, in Ten Volumes*, vol. 8, *Soldier Life/Secret Service*. New York: Review of Reviews.

Miller, John. 2012. "Communication and Innovation in the American Civil War: Comparison of Union and Confederate Implementation of Telegraph Technology." Paper presented at *Astride Two Ages: Technology and the Civil War*, Smithsonian Institution Civil War Sesquicentennial Symposium, Washington, D.C., 9–11 November 2012.

Plum, William R. 1882. *The Military Telegraph during the Civil War in the United States, with an Exposition of Ancient and Modern Means of Communication, and of the Federal and Confederate Cipher Systems*. 2 vols. Chicago: Jansen, McClurg. [Vol. 1 reprinted Cornell University Libraries Digital Collections, Ithaca, N.Y., 2010. Vol. 2 reprinted Nabu Public Domain Reprints, LaVergne, Tenn., 2010.]

Rifkind, Jarrod. 2010. *A Historical Examination of the Confederate Post Office Department: The Early Years*. Atlanta: The Sam Nunn School, Georgia Institute of Technology.

Scaife, William R. 2004. *Confederate Strategy Command and Control*. Kennesaw, Ga.: Kennesaw Mountain Historical Association.

Sears, Stephen W. 1999. "Foreword." In *Maps and Mapmakers of the Civil War*. Earl B. McElfresh, pp. 6–8. New York: Harry N. Abrams.

Sherman, William T. 1875. *Memoirs of Gen. William T. Sherman*. 2 vols. New York: D. Appleton. Reprinted New York: Da Capo, 1984.

Spears, Joseph F. 1947. The Flying Telegraph. *National Geographic*, (April):531–554.

Walker, James. 2012. *The Battle of Chickamauga, 1863*. Painting reproduced for the "Official Map and Guide, Chickamauga and Chattanooga National Military Park." Washington, D.C.: National Park Service.

War Department. 1864. *1864 Field Artillery Tactics: Organization, Equipment, and Field Services*. Washington, D.C.: Government Printing Office. Reprinted Mechanicsburg, Pa.: Stackpole, 2005.

———. 1880–1901. *The War of the Rebellion: A Compilation of the Official Records of the Union and Confederate Armies*. 128 vols. Washington, D.C.: Government Printing Office.

Zeller, Bob. 2005. *The Blue and Gray in Black and White: A History of Civil War Photography*. Westport, Conn.: Praeger.

FIVE

VETERINARY CARE
IN THE UNION CAVALRY

David J. Gerleman

From a rain-swept army camp in 1863, Capt. Charles Francis Adams recounted for his family some of the hardships of the cavalry service, noting that the men survived well enough,

> but the horses! Such a collection of crow's bait the eye of man never saw . . . they stand without shelter, fetlock deep in slush and mud, without a blanket among them, and there they must stand—poor beasts. (Ford 1920, vol. 2:101)

Such dismal treatment and suffering were by no means unusual during the Civil War, and memories of such instances were likely foremost in the mind of a later veterinary historian who charged that "in all wars, big and little, fought by the United States, poor horse management has been an outstanding blot on our military efficiency" (Merillat and Campbell 1935:75).

The subjects of horse-related issues such as care, shoeing, abuse, and veterinary treatment have been only haphazardly touched upon in existing Civil War cavalry historiography. Although the federal government spent approximately $100 million (in 1860 dollars) on horseflesh during the conflict and eventually even created a cavalry bureau to address the specialized needs of the service, it oftentimes did not get fully serviceable animals because of a cumbersome purchase system, ignorant inspectors, dishonest contractors, and a high horse mortality rate caused in part by a lack of trained veterinarians.

Since its inception, the U.S. Army had been notably lax about veterinary care and was frequently criticized throughout the nineteenth century for being "the only one [army] in the civilized world without educated veterinarians" (Bustead

1863; Anonymous 1864b:169–170). The small prewar cavalry force virtually ignored the need for veterinary surgeons and only in 1853 were serious proposals for a veterinary corps submitted to Congress, supported by arguments that educated veterinary personnel would be less expensive than continual remount purchases (Stewart 1983; Wilkes 17 and 31 January 1863). Similar advice was repeated three years later when Capt. George B. McClellan (1857:281) returned from an official observatory European tour advocating in his report establishment of farrier, cavalry, and veterinary training schools, but again Congress ignored the suggestion.

Government parsimony was not the only reason why interest in establishing a formal system of army veterinary care fell by the wayside. Before the war the American veterinary profession received scant respect with the soubriquet "horse and cattle doctor" only tauntingly applied. Attempts to increase public awareness of the need for scientific qualification in veterinary medicine met with only limited success (Anon. 1864c:22; Wilkes 21 December 1861:254; Merillat and Campbell 1935, vol. 2:131, 133, vol. 3:173, 182–183). If the outbreak of the Civil War found the U.S. Army's medical department ill equipped to deal with the flood of wounded men it soon faced, there was at least an existing skeletal administrative structure upon which to build. But the army was in no way prepared—or had even considered—what to do with the thousands of sick, wounded, and diseased horses and mules that would also unavoidably follow in the wake of any military campaign (Figure 5.1). Furthermore, the decision to field a large volunteer cavalry force inevitably meant the problem of animal wastage would be exacerbated beyond anything the army had ever previously encountered. Like so many of the other complex organizational problems arising from the war, the question of animal healthcare and treatment would be tackled on an ad-hoc emergency basis with varying degrees of success (Stoneman 1861).

Although addressing the healthcare needs of army animals scarcely had been given an official thought in the summer of 1861, a small number of veterinarians did ask the War Department "what provision the Govt. has made for the medical & surgical care of its cavalry horses & what remuneration is made if competent persons are accepted as Veterinary Surgeons?" (Hopkins 1861; Scott 1861). George Dadd, John P. Turner, J. N. Collins, Felix Vogeli, and John Bustead were among the most actively persistent advocates calling for creation of a permanent army veterinary corps, and President Lincoln (Figure 5.2), his cabinet secretaries, and other administration officials all periodically received letters from them asking for veterinary surgeons appointments.

Other applicants even made extravagant claims as to their abilities to cure a multiple of equine diseases, and nearly every aspirant made haste to refer to their

Figure 5.1. "A Cavalry Charge," drawing by Edwin Forbes (ca. 1876). Fallen, dead, and wounded men and horses in the foreground illustrate battlefield carnage; but just as for human soldiers, the numbers of horses killed in Civil War battles were vastly outweighed by deaths in camp caused by disease, malnutrition, and exhaustion. (Library of Congress, Prints and Photographs Division, Washington, D.C., LC-DIG-ppmsca-20758.)

congressional endorsements, acquaintanceships with prominent leaders, or European educations to try and win acceptance of their proposals (Florence 1861; Fuller 1861; Shearer 1861; Vogeli 1864). "On behalf of that noble animal the Horse," Dr. John N. Collins (1861) wrote Lincoln,

> I appeal to the head of our government to make a provision if it is
> not made already, for the medical treatment of our cavalry and other
> horses not alone the wounds received on the battlefield, but the many
> diseases they are liable to contract from the large number together
> and by having a Veterinary Surgeon attached to each regiment many
> thousands of dollars worth of those animals could be saved.

Similar like-minded individuals wrote to Quartermaster General Montgomery C. Meigs (Figure 5.3), pointing out that although the army reportedly had more than 40,000 animals, no provisions existed for their health and treatment. Meigs, to whom all veterinary inquiries were forwarded, responded that he recognized the importance of veterinarians, but no law existed permitting their appointment (Bustead 1861a, b; Meigs 1861).

Figure 5.2. President Abraham Lincoln, photographed by Alexander Gardner, 9 August 1863. Contrary to legend, Lincoln never offered lieutenant commissions to induce trained veterinarians to join the army as no law permitted it. (Library of Congress, Prints and Photographs Division, Washington, D.C., LC-USZ62-2279.)

To correct this lack of legislation George H. Dadd, a prominent veterinary practitioner and teacher—he was referred to by several of his prominent recommenders as "the most capable veterinary surgeon in the United States"—implored several senators and congressmen to sponsor legislation to provide the army with veterinary surgeons as a matter of national importance. Dadd argued that

DAVID J. GERLEMAN

Figure 5.3. Quartermaster General Montgomery C. Meigs, photographed in 1865 by Frederick Gutekunst. While acutely aware of the need for trained veterinarians to stem the wastage of army animals, Meigs was unable to convince Congress to create an army veterinary corps. (Library of Congress, Prints and Photographs Division, Washington, D.C., LC-DIG-ppmsca-07784.)

Congress should allow trained veterinarians to rank alongside medical staff so that they might do for the United States what had been done for England and France in preventing unnecessary death and disease in both army and civilian livestock. He also pointed out that army farriers possessed no medical training and thus their services were only of limited value (Dadd 1861a; Moore 1861). As a stop-gap measure he suggested that regiments be allowed to appoint veterinarians who then would select intelligent farriers from each company and teach them basic medical knowledge. Furthermore, Dadd advocated veterinary surgeons being given authority to establish camps for the treatment of sick animals and to issue orders to ensure camp sanitation, estimating that the expenses involved would be offset by an anticipated return to service of some 70% of broken-down animals (Dadd 1861b).

Calls even emanated to create a Veterinary Surgeon General's Office to oversee all unserviceable government horses, yet candidates for the position were told that no law allowed such an office and that veterinarians could be hired only as civilians at quartermaster depots for limited pay (Sibley 1862; Meigs 1863c). Indeed it was the lack of commissioned officer status and low wages that was the largest hurdle keeping the nation's small pool of trained veterinarians out of the army. Repeatedly, Meigs had to tell applicants that they would neither rank as army officers nor receive pay and allowances as such. Any veterinarians the quartermaster did hire would only serve for limited periods to undertake special assignments, such as conducting unit or depot inspections, and these men would be classified as civilian employees at varying levels of pay (Meigs. 1862c, 1863f).

Due to such dismal prospects it was no wonder then that professionally trained men

> could not be obtained, and those who are competent . . . are not
> willing for the very small amount of pay offered by the Govern-
> ment to serve, hence but a very small amount of benefit has been
> derived . . . to what should have been, had competent veterinary
> surgeons been employed, or sufficient inducement been offered to
> bring them into the field. (Anon. 1864a:6)

(See also Wilkes 16 May 1863:163; Merillat and Campbell 1935, vol. 2:147–154; Madison 1879:176)

Army and congressional reluctance to form a permanent veterinary corps arose from the cost-saving and seemingly ingrained belief that company farriers possessed veterinary knowledge and therefore could act as capable horse doctors. The results were generally not encouraging. Thomas Agan was a "fine fellow" appointed to be horse doctor for his company, but by all accounts he knew less about horses than any other man in the unit (Goodhart 1896:80–81). The situation was similar even among the Regulars; in Sidney Davis's company, the black-smith served as its horse doctor, but "as a veterinary surgeon I do not think he was successful in many instances, though he would stand for hours grandly contem-plating his patients . . . giving preference to those men who wanted their horses shod." The same unit had little better success when it employed a civilian whose only claim to distinction, Davis added, was his ability "to look wise and say noth-ing" (Davis 1994:54, 338; see also Fry 1862:348).

Such medical mismanagement dismayed army horse lovers; one officer was disgusted to find that

> the doctoring of sick horses is no one's particular business, and as
> a general thing all think that they are competent . . . to administer.
> Such practice combined with disease soon finishes the work, and
> the horse is no more; what is everyone's business is no one's. Mal-
> practice in horse is just as fatal as in man . . . for medicine, of all
> things, injudiciously used or used by the uninformed, are more fatal
> than effectual. (Anon. 1864a:5)

DAVID J. GERLEMAN

Figure 5.4. Camp of the 18th Pennsylvania Cavalry, near Brandy Station, Va., in March 1864. Many farm-raised troopers employed home cures to heal their mounts or appealed to their own regimental surgeons to extract bullets or sew up wounds. (Library of Congress, Prints and Photographs Division, Washington, D.C., LC-B8171-7625.)

Even worse, ignorant diagnoses meant that potentially curable maladies like distemper were misdiagnosed as glanders, resulting in many horses being needlessly shot (Turner 1862; Meigs 1864d:889; Denison 1876:77).

The lack of trained horse doctors forced many officers and men to look after their horses themselves (Figure 5.4), with officers, like Charles Russell Lowell, perhaps being more solicitous owing to their vested financial interest and means to employ servants (Emerson 1907:327, 338). Some troopers, like those of the 3rd Pennsylvania, took the utmost pains to heal their mounts and "had rather injure themselves than have their horses harmed," preferring to nurse their favorite wounded animals rather than request replacements "as their owners would not part with them." Having to tend to ailing livestock was nothing new to farm boys amidst the rank-and-file who no doubt often harkened back to home cures to try mend their cavalry chargers, such as treating horses' fevered leg joints and sore heels by applying a variety of poultices (Glazier 1870:131; Regimental History Committee 1905:226; Dornblaser 1884:60).

Some soldier ministrations, however, bordered on butchery; Trooper Luman Tenny employed copious bleeding to cure his horse's "belly-ache," and even Robert E. Lee used that time-honored antidote on his horse, Richmond, although the bleeding, coupled with a purgative, resulted in the animal's death within hours (Tenney 1914:35; Dowdey and Manarin 1961:243). Proof of disastrous results from allowing troopers to randomly treat mounts was well illustrated by an incident in the 3rd Pennsylvania Cavalry where a detail of farm boys was selected to act as regimental horse doctors in the hope that their rural upbringing would prove advantageous. The men were given a horse medicine chest, a book on horse ailments, and 110 diseased horses to treat. Rather than refuse the assignment, the men went ahead and formed boluses from a paste of flour, arsenic, and other drugs, orally administered them, and then waited to see the result. The next day most of the dosed horses were dead and those still alive were on their knees. More deaths followed, and the men's novice veterinary careers immediately ended (Regimental History Committee 1905:532).

Another alternative employed by troopers was to consult their own regimental doctors about how to treat their horses, extract bullets, or sew up wounds. Surgeon Elias Beck (1931:151) of 3rd Indiana extracted a bullet from his own mount that had lodged near the hip joint, and after the surgery his mount was "fattening up—full of life & as good a horse as is in the Regt." Men of the 1st Massachusetts regularly consulted Surgeon James Holland with any horse health questions—agricultural journal advice columns often advised readers to consult family physicians if no veterinarian was locally available. Holland's advice was "always good" (Crowninshield 1891:102).) He had no patience for self-proclaimed horse doctors and was outraged to discover one of them using violent remedies to treat a mare for colic when she was actually suffering from the onset of labor (Crowninshield 1891:294). When the commander of the 1st New York Mounted Rifles appealed for the appointment of an experienced and educated veterinarian, "one fully equal to the responsibilities of this most important post," he received the oft-repeated official response (Bowen 1900:311; Dodge 1863).

Meigs was aware of the need for good veterinarians, but his hands were tied by regulations and lack of congressional authority. To another officer's complaint about quack veterinary surgeons employed in West Virginia, Meigs (1863a) wrote,

> They would probably apply the same term to anyone . . . [recommended] to take their places. No person should be employed as a

Veterinary Surgeon unless skillful. It is not, however[,] probable that regular graduates of Veterinary colleges can be found in this country in sufficient numbers to supply the number needed in the present military establishment.

(See also Wilkes 25 February 1865:403; Holcombe 1881:340.)

While acceptance of trained veterinarians was largely shunned by the army as unnecessary, there were a few hesitant steps in the direction of improving animal healthcare. In May 1861 Congress added a new regular 6th Cavalry regiment, which was authorized to enlist a veterinary sergeant for each of the regiment's three battalions to be paid $17 per month (AGO 1862). Not until 3 March 1863, however, did Congress finally sanction one veterinary surgeon for each volunteer cavalry regiment to serve at the rank of sergeant-major and to be paid $75 per month (AGO 1863a; O'Brien and Diefendorf 1864, vol. 1:327–328). The act changed little since few professional veterinary surgeons could be induced to enter the army at such a lowly rank.

The army veterinary selection process was further formalized a few months after formation of an official Cavalry Bureau in July 1863. Under the new rules, veterinary surgeons were to be nominated by cavalry regiment commanders, then recommended by a regimental board of three ranking officers, and then forwarded on to the chief of the Cavalry Bureau, who finally submitted it to the Secretary of War for appointment. A record of the appointments kept by the Adjutant-General's Office shows that between 1863 and 1865 there were approximately 98 applications for appointment as veterinary surgeon, many made during the war's last year. Most of these applicants were enlisted men or farriers, with the largest number from New York, Indiana, and Pennsylvania regiments (AGO 1863c; NARA RG 108b; Card 1865b). Even this system was open to abuse. No set standards existed to qualify candidates for the post, and selection largely rested in the colonel's hands. Some cavalry commanders used the veterinary surgeon position as a berth for disabled or favored soldiers. Merit and competent performance were occasionally rewarded, such as in the case of the Chief Farrier of the 1st Vermont who was recommended to be mustered in as veterinary surgeon because he was both capable and qualified and had a proven track record of nursing many of the regiment's horses back to health (First Vermont 1864).

Meigs repeatedly had to explain to applicants how the process worked and that the quartermaster could only appoint veterinary surgeons for local depots; field service required direct application to commanders of units that needed them (Meigs 1863c; Sibley 1863a). The law raised other questions: Were the veterinary surgeons hired by the Quartermaster Department or employed by mounted artillery, for instance, still to be paid by previous arrangements (Meigs 1863d,e)? Even in 1864 there was a lack of clarity on this issue, as Gen. Philip Sheridan had to query Washington whether or not the commander of his horse artillery could legally employ a veterinary surgeon. Meigs (1864c) responded,

> the employment of veterinary surgeon[s] in the artillery arm is not authorized by existing laws. The necessity of their employment with a Brigade of Horse Artillery is as great, it would seem, as with the same number of cavalry horses and there would be no impropriety in detailing a competent non commissioned officer for such duty, but he would not be entitled to extra pay.

In 1862, the Quartermaster Department looked into supplying each cavalry regiment with a portable veterinary medicine chest to be constructed of lightweight wood with compartments large enough to hold a surgical kit and a three-month supply of medicine. Total list price for the veterinary chest complete with medication was $105.67 (Meigs 1862a; Miller 1980). It is uncertain how many cavalry units actually received the transportable chests, although George F. Parry (1864), a trained veterinarian serving with the 7th Pennsylvania, was delighted to receive in the spring of 1864 "a new Veterinary Med. Chest complete with a splendid lot of medicines." Even so, items missing from the approved army list of horse medicaments included disinfectants and clinical-diagnostic aids such as thermometers or stethoscopes. A veterinary manual was drafted "to induce men, by informing them of the symptoms, to treat more tenderly the timid life which is disposed to serve them, and is willing to love them" (Turner 1863). Submitted to the army in 1863 for possible inclusion with the horse medicine chests, it was apparently never printed or distributed (Bustead 1864; Duffey 1861).

Treating horses with armies at the front was only part of the problem; an ever-growing issue was what to do about the broken-down animals constantly being generated by the entire military establishment (Figure 5.5). Like other valuable pieces of army equipment, these animals needed to be repaired and returned to service. Early in the war the army's original method involved hiring civilian hostlers and renting pastureland so that unworked horses would, in effect, restore themselves. This method garnered only limited results since a large percentage of pastured horses and mules suffered differing diseases and often died from being indiscriminately mixed together with little oversight. Experienced cavalry leaders remained adamant that stabling under close supervision was superior as

Figure 5.5. The return of Gen. August. V. Kautz's exhausted cavalry, sketched by William Waud from an 1864 raid in Virginia. Every Civil War campaign produced a wave of sick, wounded, and exhausted horses and mules. Like other complex problems arising during the conflict, animal healthcare and medical treatment would be addressed on an ad-hoc basis with only a limited degree of success. (Library of Congress, Prints and Photographs Division, Washington, D.C., LC-USZ62-138110.)

a recruitment measure (Fosses 1864; Wilson 1865; Meigs 1864a,b; Message to Cavalry Bureau 1864; Ekin 1864a).

In July 1863 the War Department established the Cavalry Bureau—the only branch-specific agency created during the war—and installed Gen. George H. Stoneman as chief to tackle the immense task of maintaining and equipping the cavalry (AGO 1863b; Anon. 1863b:3; Kautz n.d.). One of the bureau's first acts was to establish remount-collection depots throughout the North, the largest being the Giesboro Point facility across the Anacostia River from Washington (Stoneman 1863b) (Figure 5.6). Construction at Giesboro proceeded at a feverish pace to accommodate the 17,000 unserviceable cavalry horses requiring care and treatment that autumn. It eventually burgeoned into a massive complex covering 625 acres with stables, forage houses, storehouses, blacksmith shops, barracks, mess halls, and stockyards able to hold 30,000 animals. The complex dwarfed any previous establishment for recruiting army horses and cost the correspondingly large sum of $1,225,000; monthly operating costs hovered near $180,000 (Sawtelle 1865; Rhodes 1979:328; Thomas 1863; Browning 1865; Pay Invoices 1864; Anon. 1863a:286; Ludington 1864; Stoneman 1863c; Dupuy 1864).

Figure 5.6. Corrals at the U.S. Cavalry Bureau's Giesboro Point Depot, photographed in 1864 by Andrew J. Russell approximately a year after the depot's construction began in late summer 1863. (Library of Congress, Prints and Photographs Division, Washington, D.C., LC-DIG-ppmsca-08283.)

Finding competent veterinary surgeons to check horse mortality at quartermaster and Cavalry Bureau depots was just as problematic as supplying them to the armies owing to what General Stoneman (1863a) ruefully admitted was the "the compensation now allowed by the Government." There was, however, no shortage of individuals who promised miracle cures for a wide range of horseflesh ills, although few ever achieved positive results when allowed to test their remedies (Gregson 1864; Wilkes 28 March 1863:50, 58; 29 July 1865; 19 August 1865; Merillat and Campbell 1935, vol. 2:172; *Country Gentleman* 1863; Bustead 1863; Paaren 1865:53; Chambliss 1865:582; Wilson 1864; Bramhall 1864). One such individual was Alexander Dunbar, a self-proclaimed horse foot expert, who persistently offered, if paid handsomely, to train army farriers in his secret technique to prevent lameness. He demanded a 10-year secrecy oath and a $100,000 fee. Dunbar's importunities were declined, but other quacks and snake-oil peddlers continually pestered the bureau (Grant 1988 [1866]; Merillat and Campbell 1935, vol. 1:158).

Even men initially thought competent when first employed ended up being dismissed when their treatment methods proved unsatisfactory. When John P. Turner, who had long importuned for a chance to treat army horses, was finally given the chance he turned out to be no more skilled than anyone else at the depot. He was reportedly a poor apothecary who used immense quantities of expensive medicines that killed more horses than were cured (Tompkins 1863; Meigs 1863b). The demand of would-be veterinarians to be supplied with "almost every drug and poison known to medical science" was one of the Quartermaster General's greatest concerns, and depot chiefs were warned to "have an eye to the matter that the horses are not dosed to death" (Sibley 1863b; Anon. 1863c).

Owing to the imprecise nature of the records, it is hard to state with certainty the exact numbers of animals recruited and returned to duty. Early reports suggested that roughly 50% of all unserviceable animals were eventually recruited, but during the war's last 14 months that percentage increased. Between April and September of 1864 over 12,405 sick and broken-down animals were turned into the Giesboro depot, approximately 60% of which were once again made serviceable in 30 to 90 days (Ekin 1864b; Norvell 1911:365).

All told, to fight and win the Civil War the Union fielded approximately 258 mounted regiments and 170 independent cavalry companies. Hidden within those statistics was an immense expense in both monetary terms and in lives of animals ruined or destroyed. The best estimates place the total numbers of horses procured for Union armies at over 650,000, with 75,000 animals captured in enemy territory; Rhodes (1979) listed Union appropriations for horse purchases at $123,864,915 and 825,766 horses issued to the armies. For the cavalry alone between 1 May 1861 and 1 January 1864, nearly 260,000 cavalry horses were purchased; the following year an additional 193,388 head were added to the total (Ekin 1866; Ingalls 1865:252–253; Secretary of War 1862–1863:72–73; First Division 1865). Such numbers represented an enormous military investment that even by extremely conservative estimates tallied over $95 million (Meigs 1864d:906; 1865:221).

As the war ground to a close in 1865 and Union armies demobilized, the need for veterinarians was still keenly felt. Between 1 May 1865 and 2 August 1866, over 207,000 horses and mules were auctioned off in a massive government clearance sale that was a boon to buyers, although fears abounded that the dispersal of army stock would spread deadly contagious diseases far and wide (Meigs 1865:221, 258; Stanton 1866:1032; NARA RG 92j, vol. 1:90–91, 100–102, 103–104, 126–130; Card 1865a). Farmers claiming to have been victimized by buying sick army animals cautioned others and some even believed "it [would be] better for the country if the [Quartermaster] department had killed every mule and horse owned by the United States, than to spread [glanders] all over the land" (Anon. 1865a:301–302). Leading agricultural journals, such as *Prairie Farmer* and *Wilkes' Spirit of the Times*, advised that government sales be avoided as the risk of epizootic was too great to rationalize the purchase of questionable stock. Such periodicals also reminded readers of the national need for competent veterinarians and criticized "the government's parsimonious course in that respect" (Anon.

1865a:301–302; 1865b:349; Wilkes 21 March 1863:42; Merillat and Campbell 1935, vol. 2:164, 174; Cavalry Bureau 1864).

While it can be argued that the carnage of the Civil War helped modernize the medical profession and improve the way wounded soldiers were treated, no similar claim can be made for veterinary care. The army's low priority toward securing comprehensive veterinary care is reflected in the amounts expended on it during the war. For fiscal 1862 only $2,213 was spent on veterinary surgeons with an additional $8,990 for animal medicines; in 1863 the amounts rose to $16,631 and $39,292 respectively. For 1864, there were additional increases to $46,780 and $168,159, but by the end of fiscal 1865 disbursements for veterinary surgeons dropped to $28,041 and medicines to $107,522 (Meigs 1862b, 1863g, 1864d:876, 1865:252; Paaren 1865:53; Pleyel 1864). Averaged for the entire war the U.S. Army spent only $1,951 per month for veterinary surgeons and $6,750 per month for horse medicines throughout the conflict—figures that represent only a fraction of what was needed to affect a truly dramatic difference in cavalry horse health (Figure 5.7).

Figure 5.7. "Lt. King's horse," ca. 1862–1865 (photographer unknown). Although the U.S. Army spent over $100 million on horses during the conflict, it distributed only $93,665 for veterinarian pay and $323,963 for equine medicines for the entire war (Meigs 1862b, 1863g, 1864d, 1865). (Library of Congress, Prints and Photographs Division, Washington, D.C., LC-DIG-cwpb-03786.)

Proposed reform advice, such as organizing horse hospitals with segregated wards according to disease and drugs being dispensed only on written prescription, were quickly forgotten as the army shifted from war to peace (Paaren 1865). Also tossed aside were suggestions that an official veterinary college be established despite pleas that doing so would provide "a great benefit to the public service . . . [and] promote our agricultural pursuits and be useful in . . . private walks of life" (Pleyel 1864). The drastically reduced postwar army shelved any ideas regarding veterinarians. Not until 1881 were cavalry veterinary sergeants even required to be graduates of reputable veterinary schools. For 50 years after the end of the Civil War the status, training, and authority of army veterinarians remained largely frozen. Not until June 1916, with rumblings of possible United States involvement in World War I, did Congress finally authorize creation of an army Veterinary Corps. This delayed piece of legislation at last granted veterinarians the commissioned officer rank they had long sought, although advancement to the rank of major was possible only after 20 years of service (Miller 1961). Could more have been done to save large numbers of army horses lost during the war? Certainly, but distrust of a nascent veterinary profession, lack of knowledge of microbial diseases, primitive medicines, forage shortages, and poor camp sanitation coupled with inattentive care all conspired with deadly effect on Union cavalry horses. The history of Civil War veterinarians still has yet to be fully explored, but what is transparently clear is that poor veterinary care was indeed an outstanding blot on the army.

References

Abbreviations

AGO   Adjutant Generals Office

NARA   National Archives and Records Administration, Washington, D.C.

RG   Record Group

AGO. 1862. General Orders No. 91, 29 July. War Department et al. 1880–1901, ser. 3, vol. 2:270–283.

———. 1863a. General Orders No. 73, 24 March. War Department et al. 1880–1901, ser. 3, vol. 3:85–99.

———. 1863b. General Order No. 236, 28 July. War Department et al. 1880–1901, ser. 3, vol. 3:580.

———. 1863c. General Order No. 259. War Department et al. 1880–1901, ser.3, vol. 3:605–606.

Anonymous. 1863a. "Inspection of Army Horses." In *Wilkes' Spirit of the Times*, ed. George Wilkes, 3 January 1863, 286.

———. 1863b. The Cavalry Bureau. *Army and Navy Journal*, (29 Aug.):3.

————. 1863c. "The United States Army: The Transportation Bureau and Its Value." *New York Herald*, 30 October 1863.

————. 1864a. "Army Horses." *Prairie Farmer*, 2 January 1864, 5–6.

————. 1864b. "Offices and Duties of Veterinary Surgeons in the French Army." *Prairie Farmer*, 12 March 1864, 169–170.

————. 1864c. "The Importance of Veterinary Education." *Prairie Farmer*, 9 July 1864, 22.

————. 1865a. [Article title unknown.] *Prairie Farmer*, 21 October 1865, 301–302.

————. 1865b. [Article title unknown.] *Prairie Farmer*, 11 November 1865, 349.

Beck, Elias H. 1931. "Letters of a Civil War Surgeon." *Indiana Magazine of History*, 27, no. 2 (June):132–163.

Bowen, James R. 1900. *Regimental History of the First New York Dragoons*. Battle Creek, Mich.: Author.

Bramhall, E.C. 1864. Message to Cavalry Bureau Chief, 22 April. NARA RG 92f.

Browning, George. 1865. Message to James A. Ekin, 23 March. NARA RG 92c.

Bustead, John. 1861a. Message to Simon Cameron, 30 August 1861. NARA RG 92a.

————. 1861b. Message to Abraham Lincoln, 23 December. NARA RG 92a.

————. 1863. Message to Edwin M. Stanton, 16 December. NARA RG 108a.

————. 1864. Message to Edwin M. Stanton, 24 June. NARA RG 92d, vol. 1:125.

Card, B.C. 1865a. Message to Charles G. Sawtelle, 11 November 1865, NARA RG 92b, vol. 88.

————. 1865b. Message to Felix Vogeli, 11 November. NARA RG 92b, vol. 88.

Cavalry Bureau. 1864. Circular, 28 March. NARA RG 108c.

Chambliss. W.P. 1865. Message to James H. Wilson, 13 January. War Department et al. 1880–1901, ser. 1, vol. 45, pt. 2, pp. 581–583.

Collins, John N. 1861. Message to Abraham Lincoln, 9 September. NARA RG 92a.

Country Gentleman, The. 1863. [Article title unknown.] *The Country Gentleman: A Journal for the Farm, the Garden, and the Fireside*, 29 October 1863, 289.

Crowninshield, Benjamin. 1891. *A History of the First Regiment of Massachusetts Cavalry Volunteers*. Boston: Houghton Mifflin.

Dadd, George H. 1861a. Message to Simon Cameron, 10 July. NARA RG 92a.

————. 1861b. Message to John A. Gurley, 9 December. NARA RG 92a.

Davis, Sidney M. 1994. *Common Soldier—Uncommon War: Life as a Cavalryman in the Civil War*. Ed. Charles F. Cooney. Baltimore: John H. Davis Jr.

Denison, Frederic. 1876. *Sabres and Spurs: The First Regiment Rhode Island Cavalry in the Civil War 1861–1865*. Central Falls, R.I.: First Rhode Island Cavalry Veteran Association.

Dodge, Charles C. 1863. Message to George H. Stoneman, 19 February. NARA RG 393b, vol. 2.

Dornblaser, Thomas Franklin. 1884. *Sabre Strokes of the Pennsylvania Dragoons in the War of 1861–1865*. Philadelphia: Lutheran Publication Society.

Dowdey, Clifford, and Louis H. Manarin, eds. 1961. *The Wartime Papers of Robert E. Lee*. Boston: Little, Brown.

Duffey, Edward. 1861. Message to Montgomery C. Meigs, 14 October. NARA RG 92c.

Dupuy, Horatio A. 1864. Message to James A. Ekin, 25 January. NARA RG 92i.

Ekin, James A. 1864a. Message to Christopher C. Angur, 1 July. NARA RG 92h.

———. 1864b. Message to Montgomery C. Meigs, 27 October. NARA RG 92h.

———. 1866. Message to Montgomery C. Meigs, 31 January. NARA RG 92h.

Emerson, Edward W., ed. 1907. *Life and Letters of Charles Russell Lowell.* Boston: Houghton Mifflin.

First Division. 1865. Fiscal Report, 17 October. NARA RG 92h.

First Vermont (unknown). 1864. Message to Samuel Breck, 1 April. NARA RG 94.

Florence, Thomas. 1861. Message to Abraham Lincoln, 27 November. NARA RG 92a.

Ford, C. Worthington, ed. 1920. *A Cycle of Adams Letters 1861–1865.* 2 vols. Boston: Houghton Mifflin.

Fosses, Julius F. 1864. Message to Edwin M. Stanton, 10 January. NARA RG 92c.

Fry, James B. 1862. Message to W.J. Palmer, 16 August. War Department et al. 1880–1901, ser. 1, vol. 16, pt. 2, pp. 348–349.

Fuller, R.C. 1861. Message to Simon Cameron, 22 August. NARA RG 92a.

Glazier, William. 1870. *Three Years in the Federal Cavalry.* New York: R.H. Ferguson.

Goodhart, Briscoe. 1896. *History of the Independent Loudoun Virginia Rangers.* Washington, D.C.: Press of McGill and Wallace.

Grant, Ulysses S. 1988 [1866]. Message to Montgomery C. Meigs, 10 February 1866. In *The Papers of Ulysses S. Grant*, vol. 16, *1866*, ed. John Y. Simon, pp. 54–58. Carbondale: Southern Illinois University Press.

Gregson, John. 1864. Message to G.A.H. Blake, May. NARA RG 92f.

Holcombe, A. A. 1881. "Army Veterinary Medicine." *American Veterinary Review*, 5(Nov.):335–349.

Hopkins A.J. 1861. Message to Simon Cameron, 7 August. NARA RG 92a.

Ingalls, Rufus. 1865. Report of operations, 1 July 1864–30 June 1865, 28 September. War Department et al. 1890–1901, ser. 1, vol. 51, pt. 1, pp. 251–256.

Kautz, August V. (n.d.) "Reminiscences of the Civil War." Manuscript Collection, Library of Congress, Washington, D.C.

Ludington, E.H. 1864. Message to James Hardie 10 May. NARA RG 159.

McClellan, George B. 1857, "Report on the United States Cavalry." In *Report of the Secretary of War Communicating the Report of Captain George B. McClellan, One of the Officers Sent to the Seat of War in Europe, in 1855 and 1856*, pp. 277–288. Washington, D.C.: A.O.P. Nicholson.

Madison, F.C. 1879. An Epidemic among Horses in Fort Randall, Nebraska, 1856. *American Veterinary Review*, 3(August):173–179.

Meigs, Montgomery C. 1861. Message to John Bustead, 30 July. NARA RG 92b.

———. 1862a. Message to Daniel H. Rucker, 28 October. NARA RG 92c.

———. 1862b. Quartermaster's Department annual report for fiscal year ending 30 June 1862, 18 November. War Department et al. 1880–1901, ser. 3, vol. 2, pp. 785–843.

———. 1862c. Message to Felix Vogeli, 16 December. NARA RG 92b, vol. 65.

———. 1863a. Message to Edwin M. Stanton, 10 January. NARA RG 92c.

————. 1863b. Message to R. McClure, 12 January. NARA RG 92b, vol. 65.

————. 1863c. Message to Moses W. Jenks, 18 February. NARA RG 92b, vol. 66.

————. 1863d. Message to Lorenzo Thomas, 6 April. NARA RG 92b, vol. 67.

————. 1863e. Message to John Arnold, 13 May. NARA RG 92b, vol. 68.

————. 1863f. Message to Julius Stahel, 18 May. NARA RG 92b, vol. 68.

————. 1863g. Quartermaster Department annual report for fiscal year ending 30 June 1863, 4 December. War Department et al. 1880–1901, ser. 3, vol. 3, pp. 1118–1126.

————. 1864a. Message to James A. Ekin, 2 June. NARA RG 92e.

————. 1864b. Message to James A. Ekin, 2 June. NARA RG 92f.

————. 1864c. Message to Henry W. Halleck, 30 September. NARA RG 92b, vol. 80.

————. 1864d. Annual report of operations of Quartermaster Department for fiscal year ending 30 June 1864, 3 November. War Department et al. 1880–1901, ser. 3, vol. 4, pp. 874–918.

————. 1865. Annual report of operations of Quartermaster Department for fiscal year ending 30 June 1865. To Edwin M. Stanton, 8 November. War Department et al. 1880–1901, ser. 3, vol. 5, pp. 212–301.

Merillat, Louis A., and Delwin Campbell, 1935. *Veterinary Military History of the United States: With a Brief Record of the Development of Veterinary Education, Practice, Organization and Legislation.* Kansas City, Mo.: Haver-Glover Laboratories.

Message to Cavalry Bureau. 1864. 17 May 1864. NARA RG 92g.

Miller, Everett B. 1961. "Evolution of Military Veterinary Medicine, 1775–1916." In *United States Army Veterinary Service in World War II*, chap. 1, pp. 1–4. Washington, D.C.: Office of the Surgeon General, Department of the Army. http://history.amedd.army.mil/booksdocs/wwii/vetservicewwii/chapter1.htm (accessed 5 September 2012).

————. 1980. "A Veterinarian's Notes on the Civil War." Paper presented for the 3rd formal meeting of the American Veterinary Historical Society, Washington, D.C., 22 July 1980.

Moore, W.H. 1861. Message to Salmon P. Chase, 12 December. NARA RG 92a.

NARA RG 92a. Letters Received by the Secretary of War and Transferred to the Quartermaster General, 1861–1862.

NARA RG 92b. Letters Sent by the Quartermaster-General.

NARA RG 92c. Quartermaster-General, Consolidated Correspondence.

NARA RG 92d. Quartermaster-General, General Halleck's Book.

NARA RG 92e. Cavalry Bureau & 1st Division, Letters Received 1863.

NARA RG 92f. Cavalry Bureau & 1st Division, Letters Received March to April 1864.

NARA RG 92g. Cavalry Bureau & 1st Division, Letters Received 1864.

NARA RG 92h. Press Copies of Letters Sent to the Quartermaster-General 1864–69.

NARA RG 92i. Cavalry Bureau & 1st Division, Letters Received 1863.

NARA RG 92j. Quartermaster-General, Horse Purchases and Sales by Department.

NARA RG 94. Orders and Letters Books.

NARA RG 108a. Letters Received by the Cavalry Bureau.

NARA RG 108b. Applications for Appointment at Veterinary Surgeon, 1863–1865.

NARA RG 108c. Letters Sent by the Chief of the Cavalry Bureau.

NARA RG 159. Inspector General's Office, Letters Sent, 1863–1876.

NARA RG 393a. Army of the Potomac, Cavalry Corps, Letters and Telegrams Sent.

NARA RG 393b. Army of the Potomac Cavalry Corps, Letters, Telegrams, Reports, and Lists Received, 1861–1865.

Norvell, Guy S. 1911. The Equipment and Tactics of Our Cavalry 1861–65 Compared with the Present. *Journal of the Military Service Institution of the U.S.*, 49:360–376.

O'Brien, Thomas M., and Oliver Diefendorf. 1864. *General Orders of the War Department, Embracing the Years 1861, 1862 & 1863*. 2 vols. New York: Derby and Miller.

Paaren, Nicholai H. 1865. Message to George S. Browning, 12 April. NARA RG 92c, entry 225.

Parry, George F. 1864. Civil War Diaries, 5 April. George F. Parry Family Papers, Manuscript Collection, Historical Society of Pennsylvania, Philadelphia.

Pay Invoices (Construction Worker Final). 1864. May. NARA RG 92c.

Pleyel, Emanuel J. 1864. Message to James H. Wilson, 8 February. Robert Todd Lincoln Collection, Manuscript Collection, Library of Congress, Washington, D.C.

Regimental History Committee. 1905. *History of the Third Pennsylvania Cavalry in the American Civil War 1861–1865*. Philadelphia: Franklin Printing.

Rhodes, Charles D. 1979. "Mounting and Remounting of the Federal Cavalry." In *The Photographic History of the Civil War*, vol. 4, *The Cavalry*, ed. Francis Trevelyan Miller, pp. 319–336. New York: Review of Reviews. [Originally published 1911.]

Sawtelle, Charles G. 1865. Message to James A. Ekin, 25 February. NARA RG 92c.

Scott, John. 1861. Message to Simon Cameron, 2 November. NARA RG 92a.

Secretary of War. 1862–1863. *Report of the Secretary of War*. H. Exec. Doc., 37th Cong., 3rd sess.

Shearer, S.C. 1861. Message to Simon Cameron, 12 November 1861. NARA RG 92a.

Sibley, Ebenezer S. 1862. Message to Felix Vogeli, 1 December. NARA RG 92b, vol. 64.

———. 1863a. Message to J. Scott, 2 January. NARA RG 92b, vol. 65.

———. 1863b. Message W.W. Smith, 16 July. NARA RG 92b, vol. 70.

Stanton, Edwin M. 1866. Message to Andrew Johnson, 14 November. War Department et al. 1880–1901, ser. 3, vol. 5, 1031–1045.

Stewart, Miller J. 1983. "Too Little, Too Late." *Modern Veterinary Practice*, (Nov.):894–898.

Stoneman, George H. 1861. Message to Robert B. Marcy, 17 September. NARA RG 393a.

———. 1863a. Message to Edwin M. Stanton, 15 October. War Department et al. 1880–1901, ser. 3, vol. 3, pp. 884–886.

———. 1863b. Message to John C. Kelton, 30 October. War Department et al. 1880–1901, ser. 1, vol. 29, pt. 2, pp. 398–399.

———. 1863c. Message to Edwin M. Stanton, 23 December. NARA RG 108c.

Tenney, Luman H. 1914. *War Diary of Luman Harris Tenney, 1861–1865*. Cleveland, Ohio: Evangelical Publishing House.

Thomas, C.W. 1863. Message to Edwin M. Stanton, 22 September. NARA RG 92c.

Tompkins, Charles H. 1863. Message to Daniel H. Rucker, 22 May. NARA RG 92b, vol. 70.

Turner, John P. 1862. Message to Montgomery C. Meigs, 25 June. NARA RG 92b, vol. 62.

———. 1863. "Formula for the Field Practice of the United States Army for the Treatment of the Horse in the Early Stages of Disease." NARA RG 92c.

Vogeli, Felix. 1864. Message to Abraham Lincoln, 26 April. Robert Todd Lincoln Collection, Manuscript Collection, Library of Congress, Washington, D.C.

War Department et al. 1880–1901. *The War of the Rebellion: A Compilation of the Official Records of the Union and Confederate Armies.* 128 vols. Washington, D.C.: Government Printing Office.

Wilkes, George, ed. 1859–1868. *Wilkes' Spirit of the Times* [periodical]. New York.

Wilson, James H. 1864. Message to W.D. Whipple, 30 December 1864. NARA RG 92f.

———. 1865. Message to W. P. Chambliss, 25 February. War Department et al. 1880–1901, ser. 1, vol. 49, pt. 1, p. 768.

PART TWO

TECHNOLOGICAL
DREAMS

CONFEDERATE SPAR-TORPEDO BOATS

Jorit Wintjes

L ike many good stories, this one begins with a bang and a considerable amount of drama. When in the evening hours of 5 October 1863 darkness was descending on the port of Charleston, a small, cigar-shaped craft was silently slipping off its moorings. Lieutenant William T. Glassell, Confederate States Navy, had assumed command of the CSS *David*, as the unusual vessel was named, barely a fortnight before. The *David* made for the ships of the South Atlantic Blockading Squadron, which kept Charleston closed for all but the most enterprising blockade runners. Glassell's orders were clear: "You . . . will proceed to operate against the enemy's fleet . . . with a view of destroying as many of the enemy's vessels as possible" (Tucker 1863). His craft was armed with a single spar-torpedo.

Glassell went straight for what in the eyes of the Confederates defending Charleston must have been the main prize: the USS *New Ironsides*, the Union navy's biggest ironclad, a 230-foot vessel displacing over 4,100 tons that, due to its combination of heavy armor and heavy firepower, was probably the most powerful warship the Union could muster. Stationed off Charleston since February 1863, the ship had not only seen considerable action during the First Battle of Charleston Harbor and the fighting around Fort Wagner (Roberts 1999:44–75), but also it already had been the target for an attack on the night of 20 August 1863. The torpedo vessel CSS *Torch*, commanded by Capt. James Carlin, a blockade runner, had unsuccessfully tried to hit the Union ironclad with three torpedoes, each carrying 100 pounds of gunpowder (Campbell 2000:42–52). While Carlin managed to get fairly close to his target, his attempt was frustrated mainly by the utter unreliability of his vessel's engines, prompting him to report afterwards:

It was my intention to attack one of the monitors [after the failed attack on USS *New Ironsides*], but after the experience with the engine, I concluded it would be almost madness to attempt it. . . . I feel it my duty most unhesitatingly to express my condemnation of the vessel and engine for the purposes it was intended. (Carlin 1863:499)

Six weeks later, Glassell was trying to succeed where Carlin had failed. At about 9:15 PM, the *David* was noticed for the first time from the deck of *New Ironsides*, yet it was already too late; Glassell (1877:231) later estimated having been about 300 yards from his target. Before the Union ironclad could react, it was rammed on its starboard side, the torpedo going off just a minute after the Confederate vessel had been hailed (Rowan 1863:12). The explosion left *New Ironsides* rattled but still afloat and outwardly intact—closer inspection would later reveal considerable damage, though not enough to put her out of action (Bishop 1863). Even today there is still some discussion on the extent of the damage (Roberts 1999:82–83; Campbell 2000:65–66), yet the basic fact that she did not immediately leave station shows clearly that the attack had failed in its main purpose. For a few moments after the explosion, small-arms fire was directed at the Confederate torpedo boat as it slid past the Union ironclad, and eventually two armed cutters were sent out, though they failed to locate it (Rowan 1863:13; Glassell 1877; Roberts 1999:80–83; Campbell 2000:59–66). The *David* had managed to slip off into darkness again, or so it seemed.

All was not well, however, aboard the Confederate torpedo boat. The column of water thrown up by the explosion had forced water into the ship and extinguished the fires of her engine, which was furthermore jammed by iron ballast cast loose and thrown around by the shock of the explosion. The *David* had not cunningly made off into the cover of darkness again; she had simply, and rather helplessly, drifted away. With his vessel *hors-de-combat* and apparently sinking, Glassell ordered his small crew of three—pilot, engineer, and fireman—to abandon ship (Tomb 1863:21). But this was not quite the end of that night's drama: Glassell and the fireman were captured, and it was from questioning them that Admiral Dahlgren, who at the time was commanding the South Atlantic Blockading Squadron, got firsthand information about the nature of the vessel that had attacked *New Ironsides* (Dahlgren 1863b; Campbell 2000:59–65), including a rough sketch of its general layout (Figure 6.1). The pilot of the *David*, however, eventually got back aboard, picked up the engineer from the waters of the harbor, managed to restart the engine, and finally brought the *David* back to

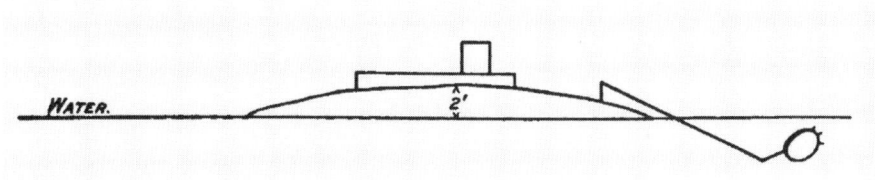

Figure 6.1. First sketch of CSS *David*, according to Rear Admiral Dahlgren (1863b:14; Public domain).

Charleston, from where she would venture out again in March 1864 in an unsuccessful attempt to attack the USS *Memphis* (Lee 1864; Patterson 1864; Tomb 1864; Campbell 2000:79–82).

Lieutenant Glassell's exploits—and even more his unusual vessel—captured the imagination of his contemporaries. In fact, after the USS *Monitor* and CSS *Virginia*, the CSS *David* may well be the most iconic warship of the American Civil War, with her dramatic attack on *New Ironsides* being one of the better-known naval engagements of the conflict. Given the novelty of torpedo warfare in general and her unusual construction in particular, it is far from surprising that interest in the *David*—and other Confederate torpedo vessels, particularly submarines—has mainly concentrated on the inventors, engineers, and men commanding these craft; their technical ingenuity; and the obstacles they faced by putting their contraptions, none of which would have passed any health-and-safety tests today, into operation.

Yet the history of the Confederate torpedo boat effort is not only a history of valiant men sailing into harm's way on a mixture of crackpot engineering and sheer bloody-mindedness; it also forms part of the overall history of the torpedo, or of the spar-torpedo, to be more precise. And whereas the later history of the torpedo already does not exactly suffer from an excess of academic interest— Edwyn Gray (2004:vii) called the torpedo "one of the world's most underresearched weapons"—the spar-torpedo has fared even worse; no modern history of the weapon itself, its employment, and its carriers is available, and general studies on the history of the torpedo either leave it out completely (Gray 2004) or devote but a few pages to it (Branfill-Cook 2014:18–20). Indeed, for an overview of the actual use of the spar-torpedo, a study published in 1880 by Charles William Sleeman (1880:187–203), who had served in the Royal Navy from 1869 to 1877 and joined the Ottoman navy afterwards, still remains the primary reference.

This chapter examines the employment of spar-torpedo boats in the American Civil War from an operational perspective and puts it into the overall context of the history of the spar-torpedo, a history that arguably begins in the United States a good half century before the war and ends before the turn of the century.

SPAR-TORPEDOES AND SPAR-TORPEDO BOATS

The CSS *David*'s October 1863 attack on the USS *New Ironsides* offers an excellent illustration of the key issues with operating spar-torpedo boats. These issues are perhaps best, if somewhat dryly, summarized by George Elliott Armstrong (1896:72), formerly Royal Navy: "In fact, the guiding principle of the spar torpedo is that its construction and design render it necessary that wherever the torpedo goes the operator must go too." It might be added that Armstrong, writing in 1896, clearly did not think the spar-torpedo to be a viable weapon anymore, as he continued:

> Nowadays it would be almost impossible for a steamboat to . . . coolly point the nose of a torpedo against her [an enemy ship's] water-line; for . . . she [the torpedo boat] would, unless the whole enemy's crew were asleep, be received with an overwhelming storm of lead and steel from the quick-firing and machine guns.

The reason for the torpedo vessel's need to get into physical contact with its target lies with the nature of the spar-torpedo itself. In principle, it was a very simple instrument: it usually comprised a long spar, 35 feet in length or more, and at its end an explosive-filled canister armed with some sort of fuse. Three factors were important for its operational success. First of all, the combination of fuse and main charge had to work—this was not the case during the *David*'s March 1864 attack on the USS *Memphis*, which was rammed twice without success, apparently causing considerable frustration among the Confederate operators (Tomb 1864; Campbell 2000:79–80). Second, the charge had to be both sufficiently powerful and—for maximum effect—placed in the right position at some point below the waterline; the lack of success during the *David*'s attack against *New Ironsides* was, on the following day, ascribed by General Beauregard (1863) either to an insufficient powder charge or to the torpedo having been placed too close to the surface. Finally, of course, with the spar-torpedo essentially being a zero-range weapon, the torpedo had to get to the target in the first place, which meant driving the spar-torpedo boat headlong into the target ship.

Whereas operating a spar-torpedo boat appears, in retrospect, to be near suicidal, one must realize that until the invention of the quick-firing gun in the 1880s and its employment on ships precisely against torpedo boats, the heavy guns of a warship usually had such a long reload time that it was, at least in theory, possible to get close to the enemy before being shot to pieces. In fact, the closer one got, the safer one actually was, as heavy guns on a warship allowed only a small degree of depression, thereby creating a "dead zone" around the ship that could not be covered by them. So having dodged the few large-caliber projectiles his enemy could hurl at him, the enterprising commander of a spar-torpedo boat would, once inside this dead zone, have to face small-arms fire only, though that of course still posed a considerable risk if his craft were detected early, as a large and alert crew was likely to open up at the torpedo boat with everything available on the ship.

Given the characteristics described above, the spar-torpedo could find employment in three different ways, all of which are in evidence in the American Civil War. First of all, a spar-torpedo could be attached to a full-fledged warship as a secondary weapon of opportunity—or as a weapon of last resort, for that matter—to increase the impact of a ramming maneuver. Perhaps the best-known example from the American Civil War is the Confederate ironclad CSS *Atlanta* mounting a spar-torpedo with a charge of 50 pounds of gunpowder (Barnes 1869:122–23; Emerson 1995:375); after her capture by Union forces on 17 June 1863, she was commissioned into the Union navy and continued to carry her spar-torpedo, though it was never actually used in action.

Spar-torpedoes could furthermore be employed on an ad-hoc basis to allow for fitting out powered boats as the opportunity for their employment arose; perhaps the most famous example from the American Civil War is the sinking of the CSS *Albemarle* in October 1864 by a spar-torpedo-armed steam launch under the command of Lt. William B. Cushing, U.S. Navy (1864; see also Cushing 1888). Cushing's craft was essentially a converted picket boat with the spar-torpedo mounted on the side of the vessel (drawings in Macomb 1864:622–623). After the Civil War and well into the last decades of the century, the spar-torpedo continued to be carried aboard U.S. warships, with the last official set of instructions published by the U.S. Torpedo Station dating from 1890 (Navy Torpedo Station 1890); an earlier version (Navy Torpedo Station 1876) also included a chapter on towing torpedoes, a concept that by the 1880s had fallen out of use. Even as late as 1890, the instructions not only made a distinction between torpedoes carried by ships and those carried by boats but also described how "torpedoes may be readily improvised from kegs or casks" (Navy Torpedo Station 1890:24–25).

Finally, in a small vessel the spar-torpedo could constitute the sole or primary armament, resulting in a "true" spar-torpedo boat. During the American Civil War such vessels were used nearly exclusively by the Confederate navy; while general circumstances prevented the building of classes of torpedo boats proper, at least three more-or-less distinct "types" of torpedo boats can be made out. The CSS *David* was built to maximize protection for its small crew and to minimize its silhouette, allowing it to approach an enemy unseen. Its design included both iron covers for the small crew of four as well as ballast tanks for lowering it as far as possible into the water, though it is not quite clear whether the original *David* already had the latter feature (Campbell 2000:56).

Given the *David's* apparent success, it is hardly surprising that she served as a model for a number of other torpedo boats built during the war. Two of these are shown in a set of pictures taken after the war in Charleston harbor, where Admiral Dahlgren reported three operational boats sunk by their Confederate crews in Charleston harbor after the fall of the city and six others in various state of (dis)repair (Dahlgren 1865:387, 402). Apart from the Charleston-based torpedo boats, at least one David-type boat was active in Mobile. It had originally been built on a government contract in Selma, Alabama, by a man called John P. Halligan, whose conduct apparently failed to impress Maj. Gen. Dabney H. Maury (1865:267), who reported in January 1865 that

> from his whole course I became convinced he had no real inten-
> tion of attacking the enemy and that the only practical purpose the
> *Saint Patrick* was serving was to keep Halligan and her crew of six
> able-bodied men from doing military duty.

Eventually, Maury effectively confiscated the boat that apparently had been lying at Mobile since June 1864 (Johnston 1864:936), its existence known to Union forces since late October (La Croix 1864; Welles 1864). The *Saint Patrick*, under the command of a Confederate navy officer, attacked the double-ended gunboat USS *Octorara* during the night of 27 January 1865, though without success as the torpedo failed to explode (Maury 1864; see also Hurlbut 1864; Jones 1864; Campbell 2000:84–86). Another David-type boat operating out of Mobile was mentioned as being destroyed in May 1864 by a boiler explosion during an attack on blockading vessels off Sand Island (von Scheliha 1868:314). While further details about this craft are unknown, it clearly cannot have been the *Saint Patrick*. The existence of this second David-type torpedo boat may lie at the roots of conflicting reports on whether *Saint Patrick* really was a David-type boat or whether it might actually have been a submersible.

While the "Davids" form the first type of Confederate torpedo boat construction, various Confederate experiments with submarines can conveniently be grouped together as the second type (Ragan 1999). Although these are outside the scope of this chapter, it should be noted that during and after the war considerable confusion existed as to whether certain Confederate torpedo craft were in fact submarines or not. To take just one example, when Admiral David D. Porter (1878:231) observed that the Davids "drowned their own people oftener than those they were in pursuit of," he was clearly thinking of submarines like the CSS *Hunley* and not of "true" David-type torpedo boats.

This leaves vessels of yet another design as the third type. These were essentially copies of the CSS *Squib*, a small, open steam launch that offered only a minimum of protection but apparently was quite fast and maneuverable (Barnes 1864; Campbell 2000:92–94). The *Squib*, together with the *David*, is probably the best-known Confederate torpedo vessel and for some reason is the only Confederate torpedo boat to appear in the collection of statistical data on "Confederate States Vessels" published by the Naval War Records Office (Anonymous 1921:267). It was led by Cmdr. Hunter Davidson, Confederate States Navy, in an attack against the USS *Minnesota* (Gansevoort et al. 1864) without, however, actually sinking the ship (Anon. 1864; Campbell 2000:95–99). The Confederates built several similar boats, of which the CSS *Scorpion*, CSS *Wasp*, and CSS *Hornet* of the James River Squadron are the least obscure (Campbell 2000:105). They took part in the Battle of Trent's Reach in January 1865, one of the last large naval engagements in the war and the only one where, at least on paper, both sides operated spar-torpedo boats (Anon. 1865; Campbell 2000:105–115), though neither side actually used them in the intended role. While in theory Squib-type boats, which were basically only steam launches, should have been easier to build than David-type boats, they suffered from the same problem the Davids did—the lack of suitable high-quality engines for speed and maneuverability.

This difficulty of providing suitable engines not only turned out to be the perennial bane of Confederate torpedo-boat building, it also points at an important characteristic of the spar-torpedo boat: while the method of bringing the spar-torpedo into contact with the opponent—that is, ramming—could appear to be archaic in the true sense of the word, in a naval conflict of the 1860s it meant the carrier vessel required an engine that was small, reliable, and capable of turning out considerable power. That was a technological requirement evidently beyond the capabilities of Confederate industry. It is well worth dwelling for a moment on the engine problems plaguing the Confederate torpedo boat effort. Obviously, when it came to torpedo boat design, Confederate engineers displayed as much ingenuity as anyone else, or more, yet turning ingenuous designs into working warships proved to be the real problem. Spar-torpedo boats were technically demanding machines; their specifications—small to the point of being stealthy, powerful engine, and high maneuverability—tested Confederate technical capabilities to the limit.

SPAR-TORPEDO BOAT OPERATIONS IN THE CIVIL WAR: NOT EXACTLY A SUCCESS STORY

Given the limitations of the technology available and the difficulties the Confederate navy had in procuring suitable material, it is hardly surprising that on the whole spar-torpedo boats were unsuccessful during the war. Ships like the CSS *David*, CSS *Hunley*, or the launch used by Lt. Cushing certainly captured the imagination of contemporaries and could in some case even enjoy limited success, yet their impact on the course of the war was negligible, particularly because the actual number of successful attacks was minimal. Whereas a list of ships damaged or sunk by Confederate "torpedoes" (Perry 1965:199–201; Schiller 2011:139–167) appears impressive at first, closer inspection reveals that nearly all ships were sunk or damaged by mines. While the Union navy lost four monitors and four other armored vessels to mines, the only ironclad sunk by a spar-torpedo boat attack during the American Civil War was the CSS *Albemarle*. Also, of six unarmored gunboats, only one, the USS *Housatonic*, fell victim to a spar-torpedo attack by the Confederate submarine *Hunley*.

On the whole, mines proved to be a much more serious obstacle, particularly during the river campaigns. Not only were they materially effective in causing the loss of a considerable number of warships and transports, they also made various countermeasures necessary, significantly slowing progress on the rivers. Additionally, mines came in a large variety of types, some of which could be produced in the field (Bell 2003:477–492). Minefields could be prepared at fairly short notice, and the mere possibility of encountering them already had a significant impact on operations. Simply put, for a modicum of effort, mines offered great returns (Steward 1866:23–25).

This was clearly not the case with spar-torpedo boats. It has already been noted that the operation of spar-torpedo boats posed considerable technical challenges. Moreover, even if one actually saw action, its value was limited. Had the CSS *David* sunk the USS *New Ironsides* on that October day of 1863, the strategic significance of its success would probably have been minimal. Success by the *David* certainly would have given the defenders of Charleston a considerable moral boost—its importance is amply illustrated by the reward set up for its sinking, which was set at $100,000 (Roberts 1999:74)—yet the defense of the port rested primarily on a formidable combination of coastal batteries, mines, and ironclad warships, making any Union foray into the harbor a potentially very dangerous undertaking. In this context, the spar-torpedo boats were both an additional deterrent against any attempt of forcing the harbor and a weapon of opportunity with which to attack individual vessels of the blockade squadron if circumstances were favorable. Even taking individual successes into account the spar-torpedo boats were never a serious threat to the blockade itself; they were far too few in number, and the Confederate navy lacked not only the technical resources for operating them in a strategically meaningful way but also anything resembling an operational doctrine. In the few cases when spar-torpedo boats actually pressed attacks home, they did it on their own; apparently no serious attempts were made at coordinating the available torpedo craft.

The general idea of coordinating attacks by small boats armed with torpedoes, however, was not exactly a new one. In 1810, more than half a century before Lt. Glassell led the CSS *David* against the USS *New Ironsides*, Robert Fulton published *Torpedo War and Submarine Explosions* to present his ideas both on the design of a torpedo craft and how it was to be used. Fulton had in preceding years tried to convince authorities in England and France of his theories on torpedo warfare. In October 1805, he even managed to blow up a ship with a container of explosives positioned below the keel of the vessel; what Fulton termed a torpedo might better be described as a drifting mine, but it should have served to silence the sceptics. As Fulton (1810:8) dryly noted: "Capt. Kingston asserted, that if a Torpedo were placed under his cabin while he was dinner, he should feel no concern for the consequence. Occular [*sic*] demonstration is the best proof for all men." Fulton (1810:10–13) also proposed the employment of anchored torpedoes (or moored mines).

Yet another idea was to actively bring the torpedo into contact with an enemy ship. The explosives container would be attached to a line that, in turn, was attached to a harpoon; the harpoon would be lodged in the hull of an enemy ship under way, and the container at the end of the line would be sucked under the hull to explode. In the case of an enemy vessel at anchor, Fulton hoped the current would do the trick (Fulton 1810:13–20). Although the idea might seem more than only a little bit impractical, Fulton (1810:15) claimed to have "harpooned a target of six feet square fifteen or twenty times, at the distance of from thirty to fifty feet, never missing."

Setting aside the questionable reliance on currents—and the slightly irritating picture of Fulton banging away at a target with a blunderbuss-turned-harpoon-gun—the overall operational concept is nevertheless quite noteworthy. For one, he actually devised an operational concept for the individual weapon system he designed. Fulton (1810:21–23) envisaged large numbers of his small boats swarming around the larger warships of a blockading force, preferably at night, and homing on their target from different directions so as to prevent the warships from concentrating their fire. He even went so far as to calculate the relative costs of an 80-gun vessel of 600 men and of 50 "torpedo boats" of 12 men each. While the warship, according to his calculation, was a massive investment at $400,000, the flotilla of boats armed according to his plans would come in at a mere $24,300, based on the assumption that it was possible to get a harpoon gun for $30 and a blunderbuss for $20 (Fulton 1810:21). Fulton seems to have clearly grasped what have been basic principles of torpedo (and later missile) boat operations ever since: comparatively low cost allowed torpedo boats to attain numerical superiority, making it possible for them to attack from different directions, thereby significantly degrading the defensive capabilities of the intended target. And like many later proponents of torpedo-boat construction, Fulton seemed to have cared little for the inherent weaknesses of his torpedo craft, which were liable to suffer considerably from any but the best sea and weather conditions.

Contemporary texts attested to the currency of the basic concept of torpedo-boat warfare formulated by Fulton at the time of the American Civil War (Steward 1866:1; Barnes 1869:38–39), but torpedo-boat warfare played no role in the Confederate effort. For several reasons, spar-torpedo boats could not be employed in the way Fulton had envisaged. On a basic level there were quite simply never enough boats available to create a force capable of striking a major blow at one or several Union warships. Given the technical problems the Confederate operators faced, it was actually no mean feat that they managed to keep some of their boats operational over a lengthy period of time. Spar-torpedo boats were complex machines that required high-quality hardware, particularly engines. Fulton's boats, in contrast, were the standard rowboats of their day; high-end technology they were not.

Even more important, however, was another issue amply illustrated by Lt. Glassell's attack on the USS *New Ironsides*. While Glassell's crew was of a size that would have made Fulton happy, closer inspection reveals a composition very different to that of a larger steam warship. In large, sea-going warships, engineers constituted as little as one-eighth the crew; the first Royal Navy ironclad, HMS *Warrior*, on commissioning had an overall complement of 645, with only 11 engineers and 75 stokers (Winton 1987:86–87). But fully half *David's* crew, one of them an engineer, was involved in running the engine. Even had it somehow been possible to construct more spar-torpedo boats, difficulties in finding enough technically qualified personnel would have precluded operating anything but a small number at the same time.

Also, beside the technical skills, spar-torpedo boat operators also needed great courage to ply their undeniably dangerous trade. According to a postwar interpreter, service in a torpedo boat called both for a "peculiar courage" and the "common attributes of true manliness" (Barnes 1869:149). The problem of finding suitable personnel for torpedo-boat operations is illustrated by an incident aboard the CSS *Gunnison*, a small, fast steamer equipped as a torpedo craft and serving with the Confederate Squadron at Mobile. The volunteer crew assigned to attack the USS *Colorado* late in 1863 proved reluctant to undertake so dangerous a mission, and as a result it was eventually called off (Greene 1863; Hitchcock 1863:631–632; Campbell 2000:102–105). Lacking a long-standing tradition of professional service, the Confederate navy had to rely on a combination of inspired individuals commanding boats they often also had designed and volunteers with possibly very different motives. This combination could on occasion produce impressive results, but on the whole it was not possible to make up for the lack of a professional *esprit de corps*. This shortcoming points at a key issue with the employment of spar-torpedo boats. Not only was a suitable industrial base needed but also a navy that had a sufficient pool of suitable manpower from which a new corps of operators of these weapons could then have been formed.

Apart from the number of boats available and the composition of their crews, the most important obstacle to a proper employment of spar-torpedo boats was the lack of a suitable operational doctrine as well as the means to put it into practice. In fact, even had there been enough Confederate boats for a massed attack, such an operation would likely have run into enormous difficulties as one of the key requirements for the success of such an operation was communication. A successful attack required at least a minimum of central control over the participating boats, something beyond the technological capabilities not only of Confederate forces but indeed of anyone else at the time. Also, the tactical understanding of how best to operate torpedo craft was extremely limited, as the Battle of Trent's Reach showed. In theory this engagement should loom large in the history of spar-torpedo boats as the only naval engagement during the American Civil War in which both sides fielded spar-torpedo boats. Yet they accomplished nothing, although the Union side may have had as many as 23 torpedoes (Lay 1865:654). Even so, the Battle of Trent's Reach clearly shows that tactically the spar-torpedo was seen as a melee weapon, much like the ram on larger ironclads. This concept failed to recognize the need for a modicum of room allowing the torpedo boat to exploit both its speed and its maneuverability, particularly if several boats were to be employed as a group.

Taken together, the issues of limited numbers of often technically deficient boats, the shortage of qualified personnel, the absence of suitable means of communication, and the general lack of understanding how best to operate these boats did not prevent tactical success in single actions, but they made operational success pretty much impossible.

SPAR-TORPEDO BOATS IN THE CIVIL WAR:
HOW FOREIGN OBSERVERS SAW THEM

The potential of spar-torpedo boats was clouded not only by the limited success they enjoyed but by the apparent ease with which effective countermeasures could be put in place. Countermeasures ranged from guard boats to devices preventing the boat from actually contacting the target ship. Admiral Dahlgren (1863a:11) ordered such measures in reaction to the attack on the USS *New Ironsides*. The torpedo threat provoked responses from Confederate as well as Union officers (Bradford 1864; Dahlgren 1864; Mitchell 1864). That these measures were quite effective is suggested by the interrogation of Confederate engineer M. M. Gray in April 1865 (Bradford 1865:413). Despite the boats' flaws, observers at the time still saw considerable promise in torpedo warfare. Thus in November 1864, Confederate Secretary of the Navy Stephen R. Mallory (1864:770) noted in a letter to Cdr. James Bulloch, one of the Confederacy's most important agents in Europe, that "the recent destruction of our ironclad Albemarle, and our own similar operations against the enemy, have attracted marked attention to torpedo boats, of which the enemy already has a fleet." He recommended the construction of further boats and proposed to use guncotton as an explosive, which apparently up to that point had not been used by the Confederates for sea mines or torpedoes (Mallory 1864:771). Likewise, in the aftermath of the *Squib*'s attack on the *Minnesota*, Capt. John S. Barnes (1864:601), commanding the Union flagship of the North Atlantic Blockading Squadron USS *Agawam*, was prompted by "the simplicity of construction [of *Squib*] and its great efficiency as a weapon of war" to provide a description including a detailed drawing and to express the hope that "several of them may be built and furnished the squadron." Gabriel J. Rains, who had been a mining specialist for the Confederates during the war and afterwards worked on a *Torpedo Book* that did not see publication until recently (Schiller 2011:13–90), was firmly convinced that "5 of these Davids will conquer any ironclad" (Schiller 2011:75), thereby already indicating the important operational aspect of concentrating numerical superiority on the enemy.

It took European observers of the war not long to follow suit. The first report by a European observer on the use of mines and torpedoes during the war followed by a year Dahlgren's June 1865 report on the defenses of Charleston. The earliest detailed account of mines and torpedoes was made by Capt. Edward Harding Steward of the Royal Engineers, who went on to command a Royal Engineers' detachment in the Zulu War of 1879. He concentrated on describing technical aspects of sea mines, spar-torpedoes, and land mines. Steward (1866:1) had served in Bermuda and Halifax throughout 1865 and gathered reports by American officers who had fought in the war. After a brief historical introduction, for which he used Fulton's 1805 experiments and his 1810 book as a starting point, though for some reason obstinately misspelling Fulton's name as "Fenton" (Steward 1866:2), he described various types of sea mines, their employment, and possible countermeasures in considerable detail before turning to what he called "motive torpedoes" (Steward 1866:19). Steward (1866:8) stressed the requirement for specialized personnel for the employment both of mines and of torpedo boats, noting that "to get the boat up to the enemy is the great point, and for this cool pluck and steady nerves are required." He also identified the lack of suitably powerful engines as a key issue in torpedo boat construction (Steward 1866:19). While Steward described in considerable detail the attack by the CSS *Squib* and the sinking of the CSS *Albemarle* by Lt. Cushing (Steward 1866:20–22), he did not comment further on the impact spar-torpedo boats might have on naval warfare, though he noted that the U.S. Navy at the time of publication continued to display interest in torpedoes and was building "some small rams" (Steward 1866:26).

Two years later, Viktor Ernst Karl Rudolf von Scheliha, a Prussian who as an engineer officer with the Confederate army had been responsible for the construction of fortifications around Mobile (Lonn 1940:243) and would in later years make important contributions to the development of the automotive torpedo (Gray 2004:26–37), published a massive *Treatise on Coast-Defence* (Figure 6.2). Von Scheliha's 326-page study covered the construction of fortifications, the employment of obstructions like block ships, torpedo warfare, and the use of electric light for coastal defense purposes. Von Scheliha (1868:300) saw great potential in the use of spar-torpedo boats, claiming it to be

> the opinion of all naval authorities that a boat by means of which torpedoes may with the greatest secrecy and safety be brought into contact with the enemy's vessel and exploded at the moment of touching without damage to the operator, will form hereafter an essential part of all judicious arrangements for coast-defence.

He went on to describe designs for "true" submarines, which he considered to be both unsafe for their operators and by and large unsuitable for their intended purpose (von Scheliha 1868:300–302), before giving a detailed account

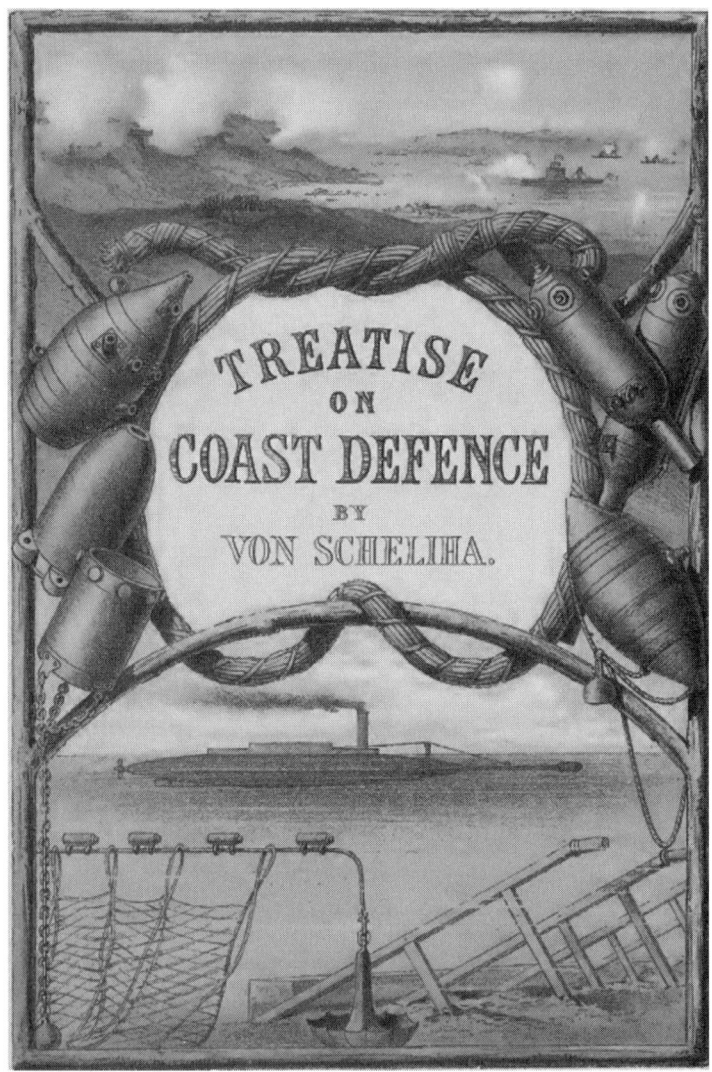

Figure 6.2. Frontispiece from Victor von Scheliha's *A Treatise on Coast-Defence* (1868; Public domain).

of Confederate torpedo-boat operations; that he even included the *Saint Patrick* should not come as a great surprise, as he—one of the engineers responsible for designing the defenses of Mobile—will have been familiar with the vessel if not with the details of its design (von Scheliha 1868:313–314).

Von Scheliha (1868:314) emphasized, just as Steward had done before him, the need of finding suitable engines for torpedo boats, stating that "the want of suitable boilers and engines may be designated as the chief reason why torpedo-boat attacks were not more frequently made from the Confederate side." For him, speed was of great importance in a torpedo boat, a requirement that effectively ruled out any oared design with which the Confederates had repeatedly experimented as well (Campbell 2000:31–41). In his eyes oared boats built in Mobile near the end of the war were merely "make-shifts" (von Scheliha 1868:315). He expected a successful torpedo boat to have a speed of at least 11 knots, an impressive figure for a power plant that was supposed to fit into a vessel roughly the size of the *David* (von Scheliha 1868:315–316)—for comparison, *New Ironsides*, though contracted for 9.5 knots, could never make more than 6.5 knots (Canney 1993:18; Roberts 1999:10), and the second-generation monitor, the USS *Passaic*, designed for 9.0 knots, could only make 7.5 on trials (Canney 1993:78).

To Steward and von Scheliha it was abundantly clear that the spar-torpedo boat had not performed well in the American Civil War for the obvious reason that it was essentially an untested weapon system poorly supported by an insufficient industrial base. At the same time, both had little doubt that rapid technical progress would eventually turn the spar-torpedo into a system playing a key role in coastal defense—rather tellingly, the frontispiece of von Scheliha's *Treatise on Coast-Defence* showed a David-type torpedo boat underway (Figure 6.2). While he had dedicated his book to Prince Adalbert of Prussia, who from 1852 onward had begun to push for the creation of a Prussian fleet, von Scheliha's real influence on the history of the spar-torpedo boat (and later the development of the automotive torpedo) was the result of his finding employment with the Russian navy after his return to Europe. The Russians had experimented with spar-torpedo boats a decade earlier, during and after the Crimean War (1854–1856); interest was renewed during the early 1860s (Fock 1979:16–17; Polmar and Noot 1991:8). With von Scheliha the Russian navy acquired important firsthand information about the Confederate use of mines and torpedoes; the combination of their own experiences and von Scheliha's knowledge eventually resulted in what arguably was the heyday of the spar-torpedo boat, its employment in the Russo–Turkish war of 1877–1878 (Sleeman 1880:192–203; Bradford 1882:31–36).

CONCLUSION

There is no denying that spar-torpedo boats played only a minor role during the American Civil War. On the Confederate side, operations were particularly hampered by the lack of suitable engines, insufficient overall technical support, and manpower requirements the Confederate forces could not readily fulfill. During the latter part of the war Union forces, while able to spend considerable resources on the construction of torpedo boats, found, as von Scheliha (1868:314) pointed out, "little opportunity . . . for making use of the torpedo in this class of offensive operations." After the Battle of Trent's Reach, the USS *Spuyten Duyvil*, easily the technically most complex spar-torpedo boat ever built (Bourne 1867:373–376; Barnes 1869:154–159; Bennett 1896:482–483), was mainly used for removing river obstacles.

Yet looking back from the last decades of the nineteenth century, many observers saw in the Confederate spar-torpedo boats the ancestors of the torpedo craft of their own day, and while few stated is as boldly as Dabney H. Maury (1894), who in an article set out to describe "How the Confederacy Changed Naval Warfare," the revolution in naval warfare that the torpedo was supposed to have brought about was commonly thought to have begun in the American Civil War. Closer observation of both the capabilities of Civil War spar-torpedo boats and the way they were employed at first seems to suggest otherwise. Not only were the vessels themselves technically handicapped, they were also mainly used for local defense or single strikes against individual targets. Later nineteenth-century thinking on the use of torpedo boats, of which the *Jeune École* is probably the best-known example, had more in common with the ideas of Robert Fulton, focusing on the employment of large numbers of torpedo craft, than with the actual operations of the CSS *David*. Indeed, Fulton's (1810:37–38) scenario of a swarm of French torpedo boats breaking a blockade by the Royal Navy probably had considerable appeal to French navy planners of the 1890s.

However unsatisfying the Confederates' experience in operating torpedo boats, their activities were nevertheless observed with great interest in Europe. As a consequence, both the spar-torpedo experiments that were undertaken by European navies during the latter half of the 1860s (Barnes 1869:190–225; Sleeman 1880:220–222; Fock 1979:17) and their employment in the Franco–Prussian War of 1870–1871 as well as the Russo–Turkish war of 1877–1878 in fact owed much to Confederate torpedo-boat operations. The Confederacy may not have changed naval warfare, but it certainly helped form the thoughts of those who did. When in 1874 the British Thornycroft shipyard delivered the spar-torpedo boat *Rap* to the Norwegian navy, a craft often seen as the first "true" torpedo boat, she was a small, sleek vessel very unlike the *David*. With a powerful engine and racy lines, she could run at more than 14 knots (Sleeman 1880:163–165, fig. 150; Armstrong 1896:165–166; Fock 1979:42) and thus conformed rather nicely to von Scheliha's ideas about how a spar-torpedo boat should look.

References

Anonymous. 1864. "Attack upon the U.S.S. Minnesota by the Confederate Torpedo Boat Squib, April 9, 1864." Naval War Records Office 1894–1922, ser. 1, vol. 9, pp. 592–604.

———. 1865. "Attempted Passage, by the Confederate Squadron, of the Obstructions in Trent's Reach, James River, January 23, 24, 1865." Naval War Records Office 1894–1922, ser. 1, vol. 11, pp. 632–694.

———.1921. "Confederate States Vessels." Naval War Records Office 1894–1922, ser. 2, vol. 1, pp. 247–272.

Armstrong, George Elliott. 1896. *Torpedoes and Torpedo Vessels*. London: George Bell and Sons.

Barnes, John S. 1864. Message to S.P. Lee, 24 May. Naval War Records Office 1894–1922, ser. 1, vol. 9, pp. 601–602.

Barnes, John Sanford. 1869. *Submarine Warfare: Offensive and Defensive*. New York: D. Van Nostrand.

Beauregard, Pierre Gustave Toutant. 1863. Message to S. Cooper, 6 October. Naval War Records Office 1894–1922, ser. 1, vol. 15, p. 20.

Bell, Jack. 2003. *Civil War Heavy Explosive Ordnance: A Guide to Large Artillery Projectiles, Torpedoes and Mines*. Denton: University of North Texas Press.

Bennett, Frank M. 1896. *The Steam Navy of the United States*. Pittsburgh, Pa.: Warren.

Bishop, T.H. 1863. Report to John A. Dahlgren, 24 November. Naval War Records Office 1894–1922, ser. 1, vol. 15, pp. 17–18.

Bourne, John. 1867. *A Treatise on the Screw Propellor, Screw Vessels and Screw Engines as Adapted for Purposes of Peace and War*. London: Longmans, Green.

Bradford, Joseph M. 1864. Message to Gideon Welles, 23 April. Naval War Records Office 1894–1922, ser. 1, vol. 15, p. 412.

———. 1865. Interrogation of M.M. Gray, 11 April. Naval War Records Office 1894–1922, ser. 1, vol. 16, pp. 411–413.

Bradford, Royal Bird. 1882. *Notes on the Spar Torpedo*. Newport, R.I.: Torpedo Station Print.

Branfill-Cook, Roger. 2014. *Torpedo. The Complete History of the World's Most Revolutionary Naval Weapon*. Barnsley, UK: Seaforth.

Campbell, R. Thomas. 2000. *Hunters of the Night. Confederate Torpedo Boats in the War between the States*. Shippensburg, Pa.: Burd Street Press.

Canney, Donald L. 1993. *The Old Steam Navy. The Ironclads 1842–1885*. Annapolis, Md.: Naval Institute Press.

Carlin, John. 1863. Report to G.T. Beauregard, 22 August. Naval War Records Office 1894–1922, ser. 1, vol. 14, pp. 498–499.

Cushing, William B. 1864. Report to D.D. Porter, 30 October. Naval War Records Office 1894–1922, ser. 1, vol. 10, pp. 11–13.

———. 1888. The Destruction of the 'Albemarle.' *Century Illustrated Monthly Magazine*, 36:432–439.

Dahlgren, John A. 1863a. Message to Gideon Welles, 7 October. Naval War Records Office 1894–1922, ser. 1, vol. 15, pp. 10–11.

———. 1863b. Report to Assistant Secretary Fox, 7 October. Naval War Records Office 1894–1922, ser. 1, vol. 15, pp. 13–15.

———. 1864. Order, 7 January. Naval War Records Office 1894–1922, ser. 1, vol. 15, pp. 226–227.

———. 1865. Report to Gideon Welles, 1 June. Naval War Records Office 1894–1922, ser. 1, vol. 16, pp. 380–403.

Emerson, William C. 1995. Unfounded Hopes: A Design Analysis of the Confederate Steamer CSS Atlanta. *Warship International*, 32:367–387.

Fock, Harald. 1979. *Schwarze Gesellen*, vol. 1, *Torpedoboote bis 1914*. Herford, Germany: Koehlers Verlagsgesellschaft.

Fulton, Robert. 1810. *Torpedo War and Submarine Explosions*. New York: William Elliot.

Gansevoort, Guert, Joseph Fyffe, and J.W. Stimson. 1864. Message to S.P. Lee, 12 April. Naval War Records Office 1894–1922, ser. 1, vol. 9, pp. 599–600.

Glassel, William T. 1877. Reminiscences of Torpedo Service in Charleston Harbor. *Southern Historical Society Papers*, 4:225–235.

Gray, Edwyn. 2004. *19th Century Torpedoes and Their Inventors*. Annapolis, Md.: Naval Institute Press.

Greene, Charles H. 1863. Message to H.H. Bell, 19 November. Naval War Records Office 1894–1922, ser. 1, vol. 20, pp. 690–691.

Hitchcock, Robert B. 1863. Memorandum to Captain Jenkins, 25 February. Naval War Records Office 1894–1922, ser. 1, vol. 19, pp. 629–633.

Hurlbut, Stephen Augustus. 1864. Message to Gideon Welles, 12 April. Naval War Records Office 1894–1922, ser. 1, vol. 21, p. 187.

Johnston, James D. 1864. Abstract log of C.S.S *Tennessee*, 16 February–31 July. Naval War Records Office 1894–1922, ser. 1, vol. 21, pp. 934–936.

Jones, Catesby ap Roger. 1864. Message to Dabney H. Maury, 16 June. Naval War Records Office 1894–1922, ser. 1, vol. 21, pp. 902–903.

La Croix, Edward. 1864. Letter to Gideon Welles, 20 November. Naval War Records Office 1894–1922, ser. 1, vol. 21, p. 748.

Lay, John L. 1865. Message to W.W.W. Wood, 8 February. Naval War Records Office 1894–1922, ser. 1, vol. 11, pp. 654–655.

Lee, Francis D. 1864. Message to Thomas Jordan, 8 March. Naval War Records Office 1894–1922, ser. 1, vol. 15, p. 358.

Lonn, Ella. 1940. *Foreigners in the Confederacy*. Chapel Hill: University of North Carolina Press.

Macomb, W.H. 1864. Abstract log of the U.S.S. *Shamrock*, 24 October–1 November. Naval War Records Office 1894–1922, ser. 1, vol. 10, pp. 620–623.

Mallory, Stephen R. 1864. Message to James D. Bulloch, 21 November. Naval War Records Office 1894–1922, ser. 2, vol. 2, pp. 769–771.

Maury, Dabney H. 1864. Message to S. Cooper, 3 February. Naval War Records Office 1894–1922, ser. 1, vol. 22, p. 269.

———. 1865. Report to S. Cooper, 26 January. Naval War Records Office 1894–1922, ser. 1, vol. 22, pp. 267–268.

———. 1894. How the Confederacy Changed Naval Warfare. *Southern Historical Society Papers*, 22:75–81.

Mitchell, John K. 1864. Message to J.R. Mallory, 21 October. Naval War Records Office 1894–1922, ser. 1, vol. 10, pp. 791–792.

Naval War Records Office. 1894–1922. *Official Records of the Union and Confederate Navies in the War of Rebellion*. 2 series, 30 vols. Washington, D.C.: Government Printing Office.

Navy Torpedo Station. 1876. *Torpedo Instructions Arranged in Two Parts: Prepared at the Torpedo Station*. Newport, R.I.: Torpedo Station Print.

———. 1890. *Spar-Torpedo Instructions for the United States Navy*. Newport, R.I.: Torpedo Station Print.

Patterson, Robert O. 1864. Message to S.C. Rowan, 6 March. Naval War Records Office 1894–1922, ser. 1, vol. 15, 35, pp. 356–357.

Perry, Milton F. 1965. *Infernal Machines: The Story of Confederate Submarine and Mine Warfare*. Baton Rouge: Louisiana State University Press.

Polmar, Norman, and Jurrien S. Noot. 1991. *Submarines of the Russian and Soviet Navies, 1718–1990*. Annapolis, Md.: Naval Institute Press.

Porter, David Dixon. 1878. Torpedo Warfare. *North American Review*, 127:213–236.

Ragan, Mark K. 1999. *Union and Confederate Submarine Warfare in the Civil War*. Cambridge, Mass.: Da Capo Press.

Roberts, William H. 1999. *USS New Ironsides in the Civil War*. Annapolis, Md.: Naval Institute Press.

Rowan, Stephen Clegg. 1863. Report to John A. Dahlgren, 6 October. Naval War Records Office 1894–1922, ser. 1, vol. 15, pp. 12–13.

Schiller, Herbert M., ed. 2011. *Confederate Torpedoes: Two Illustrated 19th Century Works, with New Appendices and Photographs*. Jefferson, S.C.: McFarland.

Sleeman, Charles William. 1880. *Torpedoes and Torpedo Warfare*. Portsmouth, UK: Griffin.

Steward, Edward Harding. 1866. Notes on the Employment of Submarine Mines (Commonly Called Torpedoes) in America during the Late Civil War. *Papers on Subjects Connected with the Duties of the Corps of Royal Engineers*, 15:1–28.

Tomb, James H. 1863. Report to J.R. Tucker, 6 October. Naval War Records Office 1894–1922, ser. 1, vol. 15, pp. 20–21.

———. 1864. Extract from notebook of First Assistant Engineer Tomb, C.S. Navy. Naval War Records Office 1894–1922, ser. 1, vol. 15, pp. 358–359.

Tucker, John R. 1863. Message to W.T. Glassell, 28 September. Naval War Records Office 1894–1922, ser. 1, vol. 15, p. 12.

von Scheliha, Viktor Ernst Karl Rudolf. 1868. *A Treatise on Coast-Defence*. London: E. and F.N. Spon.

Welles, Gideon. 1864. Message to D.G. Farragut, 1 November. Naval War Records Office 1894–1922, ser. 1, vol. 21, p. 712.

Winton, John. 1987. *Warrior: The First and the Last*. Liskeard, UK: Maritime Books.

SEVEN

ARMOR, MANHOOD, AND THE POLITICS OF MORTALITY

Sarah Jones Weicksel

On 30 October 1875, the front page news of the New Haven, Connecticut, *Columbian Register* announced a "mysterious find" in the sand along the lakeshore in Plattsburgh, New York: one-half of a bulletproof vest. The article revealed the macabre details of the steel and fabric object, its patinated surface "rusted, as if by blood"; a "deep indenture" in the metal covering the heart; traces of blood "clearly visible on its lining" (Anonymous 1875:1). The history of this "tragical relic," the newspaper declared, was "veiled in mystery," except for one clue: a business card pasted inside the steel breastplate that read, "Smith's Patent Bullet Proof Vest, made by G.D. Cook & Comp., New Haven, Conn." How the vest came to be buried in the sandy banks of Lake Champlain is indeed unclear, but the broader history behind the vest is far more tangible. It is a history of invention and failed technology; of advertising and consumerism; of cowardice, honor, and manliness. It is a history of bloodshed, war, and conflicting attitudes toward mortality and death.

The American Civil War drew millions of men from their homes and places of work and thrust them into an experience in which they confronted death and destruction on an unprecedented scale, shaking the foundation of their beliefs about manhood, duty, and death itself. In the North, the question of how to quell the rebellion with the loss of as few lives as possible weighed heavily on civilians, soldiers, and government officials alike. Inventors, manufacturers, and retailers were quick to respond to this desire to prevent battlefield deaths, resulting in the appearance of various forms of body armor. Indeed, despite the *Columbian Register*'s 1875 attempt to elicit intrigue in its readers about the rusting metal vest found in the sand, just over one decade earlier it was not uncommon to find advertisements for bulletproof vests of this same design in the pages of *Harper's Weekly*, *Frank Leslie's Illustrated Newspaper*, local newspapers, and gentlemen's clothing catalogs. In fact, the *Columbian Register* itself published news articles related to these vests.

Vests similar in design to the Smith's Patent Bullet Proof Vest were the most widely advertised and remarked upon form of body armor produced, marketed, and sold during the Civil War, primarily to moneyed and middle-class consumers. But many soldiers hesitated to accept the technology embodied by this protective gear—both its effectiveness and the implications of wearing it were suspect. The rhetoric of service to God, nation, and comrades, Drew Gilpin Faust (2008) has argued, helped men rationalize dying in the war's violence. What, then, were the cultural implications of attempting to defy death? To wear a bulletproof vest into battle?

The discourses surrounding the manufacture, advertisement, use, and, ultimately, rejection, of protective garments helped shape the shifting relationships between American conceptions of manhood and death in the wake of cultural and political turmoil over internal war, abolition, developments in weaponry, and the deadliness of the battlefield. In these objects, patriotism, moral prowess, bravery, and strength—both physical and mental—intertwined with the promises of technology. For many men, however, the use of body armor raised the specter of cowardice. Although seemingly utilitarian in nature, armor, as a garment worn upon the body, is a deeply personal object. Dress and other bodily adornments function not only as mediators between the human body and the environment but have long been regarded as a reflection of the wearer's inner self. Clothing is an outward, visible sign of a person's inward, invisible state; it is a cultural medium that shapes and communicates personal and social identity (Turner 2007:84). The technologies used to produce garments, then, inherently have cultural underpinnings.

Scholars have not probed the link between cowardice and bulletproof vests, instead assuming that mockery was a natural response to their use. But this question of cowardice and armor had far deeper implications; wearing a bulletproof vest called into question both a man's willingness to sacrifice his life for the sake of the nation and the nature of the war's effect on everyday life. In this context, cowardice, death, and manhood were entangled. Such a technological development fit uncomfortably into mid-nineteenth cultural expectations and gender norms, but the exigencies of war demanded that men and women confront the possibility for, and implications of, the use of body armor.

Personal armor was far from new in the 1860s; for centuries soldiers had employed various forms of armor for battlefield protection. Developments in ballistics technology, however, far outpaced that used in the manufacture of protective gear. As the precision and effectiveness of weaponry improved, protecting the body from harm became all the more difficult. Indeed, as Charles Ffoulkes (1912:57) observed, the disinclination to wear armor began as early as the sixteenth century. But it was with the advent and widespread use of the rifle-musket that ammunition became most deadly (Hess 2008:79). On 6 July 1861, *Scientific American* published an article articulating the use of armor in "ancient times" and noted that with "improvements in destructive agents of warfare," steel armor gradually went out of use. The article (Anon. 1861b:9) went on to assert that

> instead of abandoning mail armor with the general introduction of firearms, it would be the very thing to modify, in a great measure, the advantages of firearms, still leaving the victory to the strong man, instead of the skilled, but perhaps light, marksman.

The writer admitted that a coat of mail would not protect a man from artillery fire but believed it would be effective against bullets, noting that he "would not be surprised if steel coats would again come into use in armies" (Anon. 1861b:9). During that same year, northern manufacturers and inventors renewed efforts to design and produce appealing body armor in an attempt to profit from a burgeoning market of thousands of soldiers and their families.

Taking advantage of that potential market, however, meant not only producing saleable items but also dissociating body armor from accusations of cowardly behavior in battle. The connections drawn between armor use and cowardice reflected the nature of American conceptions of manliness, class privilege, and suspicion of the marketplace. From medieval to early modern times, the use of armor was a mark of nobility in Europe. It held a popular status, serving not only functional purposes on the battlefield but also decorative, ceremonial purposes (Sinkević 2006:14–16, 51). The use of armor was revived in France during the Napoleonic Wars, when certain elite groups of cavalry known as cuirassiers were outfitted with heavy breastplates and back plates (Bruce 2008:98). Wearing armor marked a European man with distinction—a right reserved for nobility and, in later periods, required as part of an elite fighting force. Such tradition did not, however, translate into the American context, where wearing armor was a distinctly personal decision—an attempt to evade death via mass-produced goods.

The effort to design bulletproof body armor was one element of a more widespread concern about the potential use and effectiveness of steel that pervaded discussions within both the scientific and military communities during the Civil War, particularly in regard to ships. Numerous newspaper articles appeared with titles that included "Steel Clad Steam Chariots of War" (Anon. 1861b:9) and "Iron Clad Ships of War"; indeed, a fascination with ironclads more generally took hold of the American public in the 1860s. For those who took on the challenge of manipulating steel for military protection, the "suitability of armor" posed two primary problems: it needed to be as light as possible yet still effectively withstand bullets or, in the case of ships, artillery fire. As *Scientific American* noted (Anon. 1861f:266),

> We should always endeavor to obtain the greatest strength with
> the least possible weight of metal, because a high speed is just as
> necessary to an efficient war vessel as strong sides, and every tun in
> weight saved in the armor is tantamount to an increase of speed,
> according to the model and power of engines.

Although the tests on iron and its compounds, including steel, carried out in Great Britain and the United States in the 1840s and 1850s were geared toward determining the best metal with which to encase ships to resist shell and shot, the same principles applied to bulletproof garments manufactured for soldiers (Holley 1865). Vests and other forms of body armor needed to deflect bullets while not adding excessive weight to a soldier's gear and still permitting full range of movement. In its 1861 review of the Smith bulletproof vest, which was reprinted by both northern and southern newspapers, *Scientific American* noted that the vest could "be worn with ease by any officer or soldier during the most active exercise" (Anon. 1861e:264). The inventor and manufacturer had, the writer asserted, successfully created a garment that was "very strong in proportion to its weight" (Moore 1862, vol. 3:13).

Numerous examples of armor were submitted to the U.S. Army, presented to President Abraham Lincoln, commissioned by individual soldiers, purchased from peddlers and sutlers, and sold in retail stores. In an 1890 memoir of his time serving as Lincoln's personal clerk, William Stoddard (1890:40) recalled the broad range of new inventions that passed by the president's desk, including a "pretty blue shell of polished steel" that claims to be bullet-proof, and Mr. Lincoln says that if that's the case, he approves of it; but that there must be a thorough test made. The inventor can put it on, and a detail of sharpshooters can practice at it, to see whether or not a bullet will go through.

Whether or not Stoddard accurately portrayed Lincoln's response to the invention, his anecdote captures the interplay between technology, possibility, and skepticism in regard to many new wartime inventions intended for personal use, and especially body armor.

Two primary styles of body armor survive in modern museum collections: one in the style of the Smith's Patent Bullet Proof Vest, produced by G. D. Cook and Company and also advertised as the "Soldiers' Bullet Proof Vest" (a nearly identical vest was produced by M. A. Benjamin), and the Atwater Armor Company's "Adjustable Armor" (Peterson 1973:305). These companies, both based in New Haven, led the initiative to produce, market, and sell body armor. The production of armor represented an effort to profit from wartime needs. At the outbreak of war, Cook and Company, a large carriage manufacturing firm, significantly altered its lines of production—as did many other firms. By November 1861, the company had "large contracts for carriage work for batteries, 20,000 pairs army shoes, knapsacks, haversacks, &c" (Anon. 1861h:2). To this they would soon add bulletproof vests. Both Cook and Company and Atwater Armor relied on new advances in steel technology that allowed for armor manufacture, but their design resulted in markedly different aesthetics. Cook and Company focused on steel and fabric vests, whereas the Atwater Armor Company manufactured a steel garment that shielded the entire torso and pelvis. While Atwater's "Adjustable Armor," was bulky and more akin to medieval armor, the designer of Smith's vest went to great lengths to disguise the steel plates, sheathing them in cotton and wool fabric so that, from a distance at least, they appeared to be an ordinary soldier's vest. This concealed design appealed to soldiers who desired protection but, fearful of comrades' taunts of cowardice, sought to hide their use of armor.

Vests in the collections of the Pitt Rivers Museum and the State Historical Society of Iowa illustrate the methods used to construct the Smith style of vest produced by G. D. Cook and Company. Although marketed as a bulletproof vest, the garment was actually made with two steel breastplates. According to *Scientific American* (Anon. 1861d), the breastplates were made from spring steel, known for its combination of strength and fatigue resistance. Weighing 1.7 pounds each, but

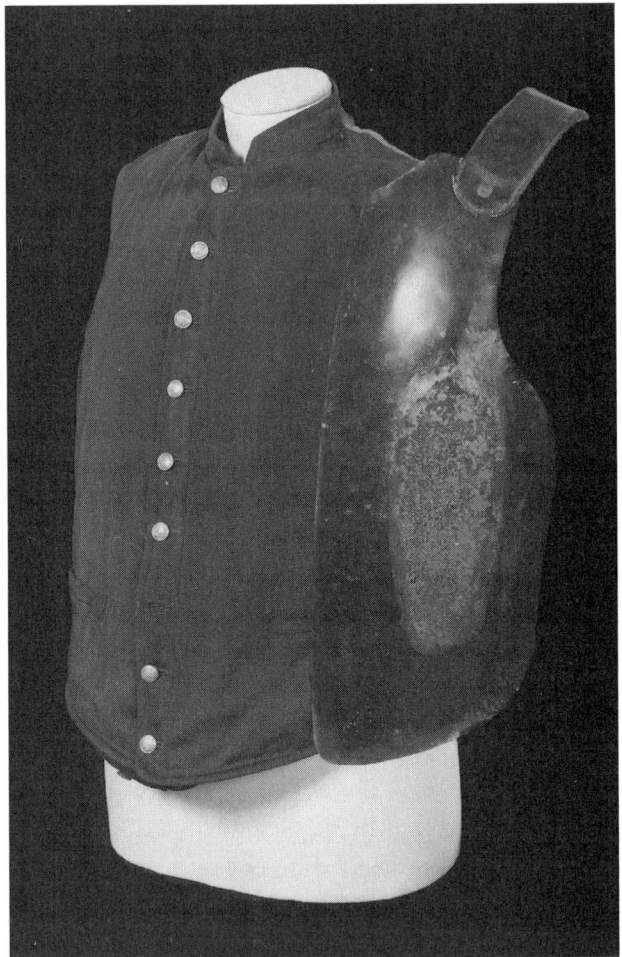

Figure 7.1. Bulletproof vest, manufactured by G. D. Cook and Company, New Haven, Conn., ca. 1861–1865. (Courtesy Pitt Rivers Museum, University of Oxford, Oxford, UK; Accession no. 1884.31.12.)

only 1 millimeter (0.04 inch) thick, the plates of the Pitt Rivers vest were also japanned to retard the formation of rust (Figure 7.1). In addition, some retailers sold a heavier version of Smith's bulletproof vest, weighing 6 to 7 pounds, for cavalry and artillery use (Stokes 1863:26). The steel was shaped to roughly fit the curvature of a man's rib cage. Each of the breastplates in the Pitt Rivers Museum vest had a leather, strap-like hook, secured by a rivet, that fit over the shoulder. These shoulder hooks, however, may have been added after the vest's manufacture in order to improve its fit; the metal strips attached to the shoulder of the Iowa vest appear to be covered in oil cloth, whereas other examples show no evidence of rivets. Alternatively, a vest advertised by Charles Stokes and Company (1863:26) in Philadelphia was "supported from the shoulders of the wearer by adjustable *Steel Springs*, which, though very light, are sufficient to resist a blow of the heaviest Sabre."

Vests of the Smith style were available in three sizes, "Nos. 1, 2, and 3," of which No. 2 was marketed as a size that "fits nearly all." The breastplates could be inserted and removed from pockets sewn into the lining of an ordinary-looking military-style vest made from wool, lined in cotton twill, and interfaced with cotton canvas. A row of buttons and a flap positioned along the bottom seam of each panel of the vest secured the breastplates, while a seam beneath the arm kept them from sliding within the lining. Such construction allowed for the breastplates to be removed and thereby the cloth more easily laundered or repaired, a nod to the demands of camp life, given that soldiers frequently mended their own clothing. It also meant that the vest could be "worn with or without the steel plates" (Anon. 1862j:2). The exterior of the Pitt Rivers vest sported brass buttons stamped with "CONNECT. SIG. REIP," the seal of Connecticut, as well as the arms of the state—three vines with fruit. Although decorative in appearance, the buttons played a critical role in the vest's function—only when the vest was buttoned did the steel plates overlap so that they would, theoretically, protect the sternum and abdomen. The back of the vest, on the other hand, was merely a cream-colored fabric, offering no protection. The vest came in two styles: a private's vest and an officer's vest, which had a more expensive fabric covering and thereby commanded a higher price, usually an additional $2.00 to $3.00. (Anon. 1862e, j).

Atwater's Adjustable Armor was of another design entirely, much more like a cuirass from earlier eras than a modern-looking garment (Figure 7.2). Unlike the Smith "Soldiers' Bullet Proof Vest," no effort was made to conceal the metal plates of this body armor. Four interlocking plates of japanned sheet iron, likely shaped in a press, made up the body of the vest, to which two optional sets of double-hinged panels could be attached to serve as hip protectors. These hinges permitted the legs to flex, while still protecting the pelvis. Straps and buckles attached to the sides of the armor allowed for it to be tightened, or "adjusted," around a man's torso. This design was clearly bulky, but the manufacturers had an important selling point: the armor could be disassembled into six single panels that could then be placed into a knapsack during a march. Sheldon Thorpe (1893:15), a sergeant in the 15th Connecticut Volunteers, noted that the Atwater Armor Company purportedly "sold over two hundred of these 'iron clad life preservers' in one day." Still, at a weight of several pounds, the armor was far more than most soldiers were willing to carry. Indeed, Thorpe also recalled that "at least fifty percent of the regiment wore away and then swore away this device."

Figure 7.2. Body armor owned by Brig. Gen. William G. LeDuc, manufactured by Atwater Armor Company, New Haven, Conn., ca. 1862. (Courtesy Minnesota Historical Society, St. Paul, Minn.)

Atwater Armor and G. D. Cook and Company were ultimately the most successful in selling their firms' merchandise, but other companies and would-be inventors also tried their hand at manipulating steel, cloth, and additional materials into garments that they claimed would stop a bullet, or at least the blow of a sword. Patent records and personal narratives illustrate efforts to produce helmets, vests, greaves, shoulder scales, and stomach protectors, as well as armor that would protect multiple soldiers at once, including large-scale portable shields and mail-clad towers pushed by steam engines. As early as 26 October 1861, *Scientific American* noted that "a great many excellent inventions in the military line have been developed since our national troubles commenced" (Anon. 1861d:266). Fewer breastplates were manufactured in the southern states, which suffered from metal shortages. As Harold Peterson (1973:305) has noted, any armor that was worn in the Confederate army was generally either captured from Union soldiers or produced by local blacksmiths. Museum collections attest to the range of experimental garments made by local craftsmen for both Northern and Southern soldiers. These pieces of body armor were likely produced on a much smaller scale, or by special order for individual men.

One such garment, worn by Hiram Morford after his enlistment in the 2nd Illinois Infantry in August 1862, is of a markedly different design and, unlike the more widely marketed styles of armor, was intended to protect both a soldier's chest and back (Anon. 1886:34). In what was likely a time-consuming production process, the maker used a sewing machine—a technology that had only recently appeared on the American market—to stitch small rectangular metal plates into pockets in the fabric, resulting in a quilted appearance (Figure 7.3). Additionally, the vest had a matching cummerbund worn around the waist to protect the pelvic area, similar in its intention to Atwater's design. Secured to the body

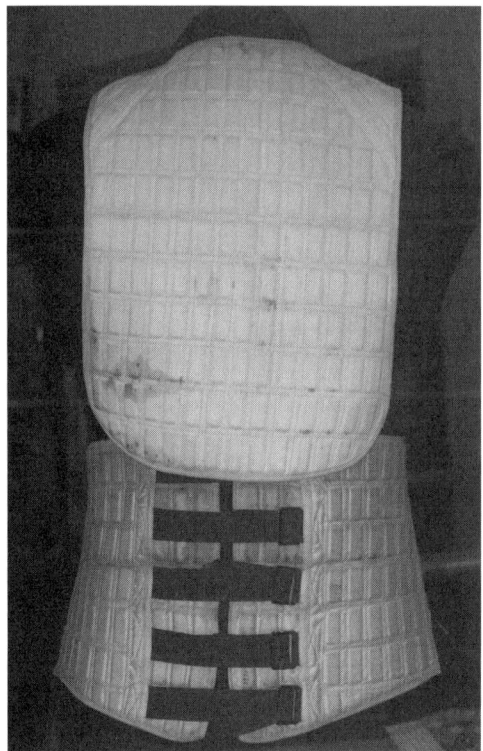

Figure 7.3. Bulletproof vest, ca. 1861–1865, owned by Hiram T. Morford. (Photo courtesy Prairie Trails Museum of Wayne County, Corydon, Iowa.)

with multiple fabric cinch straps, the garment is heavy and, while the use of many small metal plates may have allowed for greater flexibility than two large breast-plates, it was still bulky and would have been difficult to carry on the march.

MARKETING BODY ARMOR

The number of bulletproof garments sold and used during the Civil War is difficult to determine, but as Earl Hess (2008:80) has suggested, they were likely used more frequently than historians have assumed, at least initially. Indeed, Maj. Nathaniel Wales, who wore a vest of the Atwater design at the Battle of Antietam, made the more general observation that armor "was worn more often than we had any idea of" (quoted in Dean 1920:60). Soldiers' letters, memoirs, newspaper reports of battles, and surviving armor in museum and private collections support Wales's assertion.

Although not a frequent topic mentioned in personal accounts of the war, those soldiers who did discuss armor, or who were associated with its use by way of newspaper reportage, hailed from a wide range of backgrounds, among them regiments from Connecticut, Illinois, Indiana, Iowa, Massachusetts, Minnesota, New Hampshire, New Jersey, New York, Ohio, Pennsylvania, Texas, Vermont, and Wisconsin. Furthermore, written accounts and the provenances of museum objects link armor use to numerous battles that took place in Virginia, Maryland, Tennessee, and Mississippi between 1861 and 1864, and especially in 1862. Among these battles were Bull Run, Shiloh, Yorktown, Williamsburg, Corinth, Seven Pines/Fair Oaks, Port Republic, Oak Grove, Malvern Hill, South Mountain, Fredericksburg, Stones River, Gettysburg, Boonsboro, Fort Stevens, and Dandridge. The range of battles and regiments in which armor was used, along with the survival of a number of vests in varying degrees of preservation, suggests that armor use was not infrequent. Institutions holding armor and steel breastplates include the American Civil War Museum, the National Museum of Health and Medicine, the National Park Service, Pamplin Historical Park and the National Museum of the Civil War Soldier, the Minnesota Historical Society, the Pitt Rivers Museum, the Prairie Trails Museum, the Tennessee Virtual Archive, the State Historical Society of Iowa, and the Virginia Military Institute.

Multiple accounts document the use of armor at the Battle of Shiloh in April 1862. One Wisconsin officer and 26 of his men were reported to have worn armor and were "struck by rifle shot and not hurt" (Anon. 1862k:2). A breastplate found on the battlefield, now in the Tennessee Virtual Archive, further confirms the use of armor at Shiloh. Indeed, former Confederate soldier George Bryson

(1906:304) later recalled that he "saw on this battlefield [Shiloh] dead Federals having on coats made with breastplates in them." Some of these breastplates were recovered by Confederate soldiers after the battle and kept as trophies of war. New Haven's *Columbian Register* reprinted an article from the *Richmond Examiner* detailing Union soldiers' use of armor at Shiloh: "Among the trophies exhibited here are several shields taken from the bodies of men and officers, in which they were enveloped from neck to hips" (Anon. 1862g:2). Many soldiers, however, were likely buried while still wearing breastplates, their armor use therefore going unmentioned. As Drew Faust (2008:71) has shown, battlefield burial techniques focused on efficiency, often resulting in hurried burials in mass graves.

Although widely advertised and sold in northern stores and by sutlers in camp, and perhaps widely purchased, the marginal success of these bulletproof garments was short lived. Reviews of the Smith vest suggested that the manufacturer had achieved a good balance between protection and weight, but the heft of such garments remained a problem, especially for low-ranking soldiers, who carried their gear on their backs. Many volunteers initially set out from home with a host of useful yet dispensable articles, only to abandon them later. As Francis Buffum (1882:53) recalled, when all of this "paraphernalia," including "a steel plated vest," was

> arranged and mounted on the soldier's back, a giant would suc-
> cumb to such a load on a moderate march. It is not to be inferred
> that a majority of the regiment so ridiculously handicapped them-
> selves, but many did and nearly all packed their knapsacks with a
> medley which would have been judged absurd by themselves when
> settled down to genuine campaign work.

The weight of a knapsack directly correlated with a man's comfort and speed while on the march. Indeed, the ability to move soldiers quickly was so directly tied to transporting their gear that a number of U.S. Army field orders were issued requiring soldiers to abandon superfluous or nonessential belongings (NARA RG 92). Beyond tactical considerations, carrying an extra several pounds of weight was wearisome for many men, and bulky steel plates took up considerable space. As a result, the men who owned and used body armor for a length of time tended to be officers, who had better access to horse and transport wagons than did the ordinary volunteer private (Billings 1887:275).

Status and class, then, played a role in the use of this technology. Not least among the reasons for this was the sheer cost of such armor. Although the Soldiers' Bullet Proof Vest could be purchased for as little as $5.00, some sutlers sold a vest of the same style for $10.00; the more substantial Adjustable Armor purportedly retailed for nearly double the price (Peterson 1973:305). The New York City firm H.H. and Elliot offered both officers' and privates' versions of the vest, but the price was still prohibitively expensive for many men, given that $5.00 was roughly one-third of a private's monthly pay, money that many families at home depended upon for basic support (Anon. 1862d).

The availability of bulletproof garments through the emerging mass market democratized body armor. No longer a privilege of nobility, nor required by the army for uniformity, a vest could be purchased by any man who could pay for it. Given that officers were more likely to have access to bulletproof technology, both in terms of initial purchase and transport, the use of body armor held the potential to prompt discord among officers and their men, as well as accusations of cowardice. Some officers acknowledged this discrepancy, and as one Union officer attested, "many officers felt they should not be protected better than their men, consequently those who wore the armor did not advertise it" (quoted in Dean 1920:58).

The shame in advertising a man's use of body armor contrasted sharply with the shamelessness of manufacturers in advertising it. Comparing the disconnects between the effectiveness of bulletproof vests with the discourses surrounding them, as well as differences between the scientific community's perceptions versus those of military officers and soldiers, reveals a tension between technology and cultural acceptance, between military expediency and notions of manliness as they related to death. While the scientific community, including inventors and men writing for *Scientific American*, as well as some military strategists favored the development of new technologies to increase the effectiveness of the Union army, the *Philadelphia Inquirer* reported that in the summer of 1861, federal government officials had "about concluded to eschew all novelties, and stand in the beaten paths" when approached with new inventions (quoted in Anon. 1861c:73). *Scientific American*, however, urged the immediacy of developing wartime technologies, arguing that "the delay of a few days to examine and test a new invention may be of the deepest injury to the country" (Anon. 1861c:73). Among the objects tested were various forms of armor.

Proving, or at least being able to suggest with authority, just how effective these garments were was of great importance to advertisers. Articles and advertisements were worded to suggest that both the military and the federal government were involved with testing, implying that the armor had been approved by knowledgeable officials. Manufacturers appealed to the government as a regulatory body at precisely the moment during which it was under attack. Significantly, no official government entity monitored quality or safety standards, nor were there regulations regarding design, manufacturing, or testing.

Selling to a largely volunteer army with little experience in matters of war, advertisers were particularly keen to include military endorsements, a strategy to convince raw recruits of both armor's efficiency and its appropriateness. One article, for example, claimed that the armor had "been tried at Washington, and we have seen a recommendation for it from Gen. Davis, late colonel of the 104th regiment Penn. Vol." (Anon. 1862b:2). A New York City retailer advertised the bulletproof vest as "Proved and Approved by a Board of Officers at Washington" (Elliott 1862:3). Charles Stokes and Company (Stokes 1863:26) went so far as to include a report signed by Quartermaster General Montgomery C. Meigs verifying that, although experiments had not been completed, the Smith vest had not been perforated when shot with a pistol. Despite the implied official nature of such tests, the sale of bulletproof vests remained quite independent from the workings of the federal government.

The language used to market such technology reveals advertisers' anticipation of potential customers' hesitancy to purchase these garments. Attempting to preemptively assuage concerns related to both effectiveness and manliness, advertisers capitalized on wartime anxieties ranging from soldiers' desire for battlefield valor, to fears of being accused of cowardly "skulking," to home-front sentiment rooted in family members' apprehensive reading of casualty lists and desire to see their loved ones return. Bulletproof vests held the promise of protection and return. But they also raised the specter of cowardice.

CONFRONTING THE GOOD DEATH

Different, at times competing, definitions of manhood simultaneously circulate at any given historical moment (Bederman 1995:6–7). The mid-nineteenth century United States was no exception. Still, as Lorien Foote (2010:4) has argued, men shared a common goal or expectation: that others would recognize and respect their manhood. The discourse surrounding bulletproof vests lends insight into the code of manhood circulating in the popular press and amongst soldiers, and particularly officers. Skulking, "showing the white feather," to be sick with "cannon fever"—such cowardice was unbecoming of a man. Indeed for many men, unwavering duty and courage on the battlefield were central to their conception of manhood; to lack either of these qualities could stain a man's public reputation or cause him to lose favor with his comrades. As one man asserted, "Officers, as a class, must be men to whom the slightest taint of cowardice or the exhibition of fear before an enemy would be perfect destruction and everlasting indignity" (quoted in Foote 2010:100).

This fear of the "taint of cowardice" was especially culturally charged, given that dying a "Good Death" was firmly entrenched in mid-nineteenth century Protestant American society. To die a Good Death, one that showed clear signs of salvation, a person needed to outwardly demonstrate on their deathbed a consciousness of their impending death, signs of repentance, and a belief in their soul's redemption (Margaret Abruzzo, University of Alabama, "*Sin, Seduction and Scottish Moral Philosophy,*" unpublished ms). In other words, to die a Good Death, a man needed to be ready to die. This readiness to die, as Drew Faust (2008:18) has convincingly argued, was so essential to determining the "goodness" of a death that soldiers often tried to convince themselves that their comrades had been well-prepared for death, even when they died suddenly or alone. In the context of a culture in which anticipating and accepting imminent death was interpreted as an outward sign of a man's salvation, exhibiting fear, anger, or cowardliness before death, or even to have a scowl upon one's face, were distressing evidence of a man's impenitent soul. In the context of the Civil War, "dying bravely and manfully became an important part of dying well" (Faust 2008:25).

"Dying bravely and manfully" was bound up with a willingness to die for the sake of the nation. In a newspaper article calling "Men of Connecticut! TO ARMS!!," the *Connecticut Courant* (Anon. 1861a:2) expressed the widely accepted links between patriotism, manhood, and death, declaring: "It is sweet to die for one's country; and never had mortal a better cause than that which now summons all who feel themselves to be men, to rally around the flag of our fathers." This view was memorialized in the lyrics of the 1862 "Battle Hymn of the Republic," (quoted in Fahs 2001:78) imbuing patriotic death with a religious tone and righteousness:

> He has sounded forth the trumpet that shall never call retreat;
> He is sifting out the hearts of men before his judgment-seat; . . .
> As he died to make men holy, let us die to make men free,
> While God is marching on.

Many soldiers internalized this sentiment, including one man who stressed that "my first desire should be not that I might escape death but that my death should help the cause of the right to triumph" (quoted in Faust 2008:6). Officers' battlefield deaths in the summer of 1861 were venerated as redemptive national sacrifices—war heroes were symbols of nationhood (Fahs 2001:83, 86). Exhibiting cowardice in the face of battle was the ultimate mark of dishonor and disgrace, and many soldiers reiterated the phrase "death before dishonor" throughout their letters. As one soldier intimated to his mother, "It is better to die the deth of a brave Soldier than to liv a cowards life" (quoted in McPherson 1997:77).

Given the critical role played by manly resolve and sacrifice in ideas about death, to clad oneself in a bulletproof vest or another form of body armor could be interpreted as a material symbol of a man's unpreparedness to die and an unwillingness to become a martyr on the altar of the nation. In this context, wearing a bulletproof vest was most problematic when the throes of death occurred, for it was at this time that a soldier's comrades minutely scrutinized his behavior and body for possible signs of peace with God. At a moment in which the upturned corners of a dead man's mouth were interpreted as evidence of his heavenly ascent, to find a dead comrade sheathed in metal suggested precisely the opposite. A man who overtly attempted to defy death was a man who had reservations about dying; he had exhibited a desire to resist the will of God and an unwillingness to sacrifice his life for the nation. For this reason, then, many soldiers agreed with Col. Charles Johnson (2004:113) who, despite being tempted by the promises of a bulletproof vest, ultimately decided that he should place his life in God's hands and approach the enemy with bravery. Wearing body armor posed a significant cultural risk. "The best way" to confront the possibility of death in battle, Johnson explained to his wife,

> is to fully and freely place your trust in *him* who ruleth all and *face your foe* fearlessly and let him do his worst—when the order comes "pitch in" and take your chances without *skulking* behind a "*bullit proof vest*" that is not worth a continental curse at that.

The "skulking" soldier wearing a bulletproof vest was ridiculed by both his comrades and the enemy. In a letter to his wife, one Massachusetts colonel justified his decision to break his promise to inquire about armor by saying: "Bullet proofs are looked upon as indicating timidity if not cowardice. Don't think I will have one anyhow" (quoted in Samito 1998:70–71). In his memoirs, Charles De Velling (1889:119) singled out an individual soldier by name when he recounted the story of a man who bought a bulletproof vest and wore it into battle. He was wounded in the "big toe," De Velling recalled, and "the boys laughed at him considerably and suggested that he wear the vest on his feet thereafter." Indeed, it was common for soldiers to remark on the irony of a soldier being wounded in the extremities or head while wearing armor on his torso.

Similarly, Confederate newspapers took advantage of opportunities to mock armored Union soldiers, referring to the vests as "Shields for Cowards' Heart" (Anon. 1862f:2) and suggesting that wearing armor was "Very Yankeeish," and therefore cowardly. "The principle objection to the thing," North Carolina's *Fayetteville Observer* (Anon. 1861g:3) reported, "is that it gives protection to the front, whereas the Yankees most need protection to the rear." Furthermore, as William Malet (1863:187) wrote, a particular "feeling of contempt was raised among Southerners for an enemy wearing concealed armour," whether captured as a prisoner of war or found dead upon the battlefield. Cowardice was antithetical to both bravery and manliness, and in the cultural context of death, being labeled a coward had far greater implications than simply being ridiculed in camp: being a coward who was unwilling to die for your country brought the state of your soul into question.

ADVERTISING AND THE CHANGING RHETORIC OF SACRIFICE

As illustrated by soldiers' attempts to interpret even the smallest details as evidence of the surety of a man's salvation, the exigencies of war and mass, anonymous death resulted in a willingness to blur the definition of a Good Death (Faust 2008:17–18). In this context the implication of protecting oneself from a battlefield death may have been more malleable than it first appears, potentially allowing a man the cultural space in which to justify his use of protective garments. Indeed, at the same time as the soldier's death was portrayed as a patriotic sacrifice for the Union, another discourse emerged that lamented the devastating loss of young men and resented the human costs of the war. Why should men be needlessly sacrificed for the sake of honor? As one *Columbian Register* armor advertisement asked, "Do soldiers go for the purpose of being killed unnecessarily, or to put down the rebellion?" (Anon. 1862h:3).

As patriotic fervor in the early days of the war subsided, military strategy began to resemble increasingly technologically advanced modern warfare, and the deadly, bloody nature of battle became more apparent, attitudes and the way in which people approached sacrifice began to shift. By the time bulletproof vests and other forms of armor appeared on the scene in late 1861, the relationships among manhood, patriotism, death, and war were already beginning to become muddled, and a competing discourse about manly sacrifice began to emerge. Alice Fahs (2001:109) has argued that sentimental literature domesticated soldiers, describing them as boys rather than men, and reiterated that they were not professional soldiers but *citizen*-soldiers. This shift, Fahs (2001:115–16) suggests, was part of a literary process of redefining heroism so that it privileged wounded soldiers' "harder heroism of the hospital," a heroism that exceeded that of the battlefield. Being wounded, then, became a badge of honor. Such sentimental poems and stories printed in newspapers and periodicals were uncomfortably situated alongside accounts of battles, casualty lists, and advertisements for protective gear.

Through their rhetoric, advertisers were themselves crafting a particular vision of the war—one in which the relationship between the soldier and the state need not be one of martyrdom. Advertisements celebrated not the citizen-soldier whose duty it was to sacrifice his life, but the professional soldier who embraced the bravery, strength, and grit of the battlefield and whose duty it was to live to fight another day, to end the war, and to return to his home unscathed. As Gerald Linderman (1987:156–160) has shown, the culture of courage that was so prominent in 1861 was eroded over the course of the war by the overwhelming number of casualties and a shift toward the military expediency of fighting behind entrenchments. It was in this context that advertisers could simultaneously tug at the heartstrings of families and friends at home and appeal to the logic of military strategy. In this formulation, it was not only a man's patriotic, but also his familial and his manly, duty to protect his life. "There were a good many men," John Billings (1888, 275) recalled after the war,

> who were anxious to be heroes, but they were particular. They preferred to be *live* heroes. They were willing to go to war and fight as never man fought before, if they could only be insured against bodily harm.

"Iron tailors," as Billings referred to them, attempted to assuage the "situation and sufferings" of these men by manufacturing bulletproof garments. As one soldier wrote: "To be 'iron clad' when the bullets should fly as thick as hail! What more could a soldier ask?" (Walker 1909:21).

For many of these soldiers, their primary concern was the effectiveness of armored garments; others, however, needed to be assured of both the efficacy of armor *and* that wearing a bulletproof garment did not undermine their claims to manhood and patriotism. Advertisers responded by actively working to simultaneously convince potential buyers that bulletproof technology was worthwhile and distance steel garments from the taint of cowardice by suggesting that wearing armor was, in fact, the more patriotic and manly decision and that it could be worn secretly. This emphasis on both the ability to hide the use of a vest and the manliness of wearing it seems antithetical. But given the emphasis nineteenth-century society placed on the outward appearance of clothing in judging character, the ability to hide its use from others is best understood as a marketing ploy used at a time of flux in cultural attitudes toward bodily protection in battle. During a war in which men were forced en masse to confront battlefield deaths, such marketing could appeal to soldiers who believed they should protect themselves with armor yet were conflicted over how others might view them.

One particularly extensive advertisement in the *Columbian Register* (Anon. 1862h) for G. D. Cook and Company's bulletproof vest—an entire broadsheet column in length—captured the range of arguments advertisers used in marketing the use of body armor as manly and patriotic. Attempting to preserve one's life, advertisers implied, was a sign of a man's true devotion to the nation and would, therefore, enable a soldier to achieve greater valor in battle. In this view, armor-clad soldiers were more powerful and valuable to their country. They exhibited greater devotion to the nation and family than men who did not wear vests because they took additional measures to preserve their life so that they could both render more service to the national cause *and* return home to continue to support their families. The advertisement claimed that Gen. Stoneman, a well-known cavalryman, had instructed the advertiser to tell soldiers: "I wear one, for if my life is worth giving to my country, it is worth preserving for my country and family" (Anon. 1862h:3).

This appeal to the importance of the family, the protection and support of which factored into many men's understanding of manhood, extended to calls upon family members to return that protection. Interspersed among vignettes about officers saved by vests, the advertisement for G. D. Cook and Company's bulletproof vest prevailed upon fathers, mothers, sisters, wives, and friends, implying that if they cared for their soldier, they would not allow him to go into battle without the protection of iron breastplates (Anon. 1862h:3):

> Where is the father who would not furnish his son with one of
> these vests? . . . Where is the mother who would not have her son
> furnished with this truly life-preserving article? . . . Where is the
> sister who would let her brother go on to the battlefield without
> a Bullet Proof Vest? . . . Where is the wife who would not make
> almost any sacrifice to feel sure that her husband would be returned
> to her alive?

To allow a male relative to leave for war without a steel vest, then, was to fail in one's duty as a father, mother, sister, or wife. The advertisement appealed to the sentiments of home in Cook and Company's attempts to sell body armor. Lest the point be lost, the advertiser ended by prevailing upon the soldier himself, using larger, bolder type to assert the underlying point: "Every soldier should furnish himself with a Bullet Proof Vest" (Anon. 1862h:3).

In response to such advertising, friends and family members did indeed purchase and gift armor to departing soldiers. The friends of Lt. Lyon, for instance, "bestowed on him a handsome sword, sash, belt, pistol, shoulder-straps, and *steel vest.*" A sign, one of his comrades wrote, that those friends "have had an eye not only to his adornment and effectiveness, but to his safe return" (Johns 1864:53). Nathaniel Wales, on the other hand, was presented with a steel vest by his father when he departed Massachusetts but left it behind when he passed through Washington. Before the Battle of Antietam, however, Wales's "brother officer, who was wounded, insisted on my putting on his steel vest" (quoted in Dean 1920:58). Another soldier's wife was prepared to purchase a vest for him, but he informed her that he wanted only "a military vest to keep me warm, without any steel in it. I am not afraid of the bullets" (Anon. 1871:436). At least a few benefactors outfitted entire regiments with bulletproof vests, including one Massachusetts man who presented a bulletproof vest to each man in the 37th Massachusetts. Elisha Hunt Rhodes noted that the regiment was referred to as "the 'iron clads,' because when they arrived every man had a steel plate in his vest" (Rhodes 1992:112).

Advertisers also attempted to convince soldiers that although bulletproof garments might seem a new and novel technology available to soldiers, the use of steel armor was more generally becoming part of everyday life. In this way they attempted to normalize the use of what soldiers perceived to be a newfangled and therefore questionable and unproven technological item. To protect oneself with a vest, one advertisement asserted, was no different than serving on an ironclad gunboat or fighting from behind the walls of a fort—they were all simply good military tactics that employed barriers. Drawing on the names of men who had already garnered a place among brave heroes of the war, the advertisement demanded: "Does any one call Admiral Foote or Lieut. Worden cowards because they fought in iron-clad gunboats? Is Gen. Anderson any the less a HERO because he fought in a fort—(Sumter?)" (Anon. 1862h:3).

In an effort to further separate armor from accusations of cowardice, advertisers asserted that seasoned soldiers, including generals, colonels, and other officers, knew the value of protective armor and would attest to its usefulness. The advertiser, then, acknowledged outright that men would associate armor with cowardice. But to do so, they suggested, was simply evidence that a man was a new, inexperienced recruit. One article claimed that, according to Gen. Stoneman, only "men not accustomed to war will call it cowardly to wear a bullet proof vest" (Anon. 1862h:3). By asserting that veteran soldiers eagerly clad themselves in armor, advertisers appealed to Union volunteers, men who had no experience in soldiering and were conscious of their need to prove themselves in battle. Through the range of manuals describing military life, tactics, and rules, as well as newspaper articles and advertisements circulating in the popular press, new soldiers readily consumed information about how to conduct themselves upon entering the army and what gear was necessary and useful. Many feared being considered inexperienced by other soldiers, an anxiety that advertisers assuaged by assuring volunteers that bulletproof vests were only "worn by our best and bravest men," men who were experienced with the demands and conduct of war (Anon. 1863:176).

Some soldiers agreed with the rationalizations presented in advertisements, especially in regard to preserving their lives for the sake of their families. In late March 1862, for instance, Illinois officer John Cheney wrote home to his wife Mary about recent happenings in camp. "Lew Smith is here selling those bulletproof vests," he relayed. "I think they are a good thing and may buy one" (Cheney 1862–1901). Likewise, drummer Thaddeus Reynolds declared: "When I get to Elmira I am going to buy me an army steel plated vest" (quoted in Dunkelman 2004:120). Many, however, were still anxious about the potential accusations of weakness and were, as Earl Hess has noted, defensive about their decision to purchase armor. Captain William Vermillion, for instance, told his wife that he had purchased "steel armor that covers me quite well," and "I intend to wear it, not through cowardice but because I consider it my duty to protect myself in every manner possible." An intense feeling of responsibility for preserving his life had led him to wear the vest, but still, Vermillion instructed his wife: "Don't speak of it to anyone, Dollie. The boys here don't know it" (quoted in Hess 2008:81). This desire to wear protective garments, yet hide their use from one's comrades, was an element seized by advertisers in their marketing ploys.

Advertisers of the Smith style of bulletproof vest did not merely appeal to potential purchasers on an abstract level. The material aspects of the vest, including its style, also figured into their methods of persuasion. Soldiers with reservations about the potential association between armor and cowardice were assured that the garments had the "appearance precisely the same as the regular Military Vest" and that the metal plates were "entirely concealed" (Stokes 1863:26). Advertisers guaranteed that using a bulletproof vest would go undetected, going so far as to refer to the vest as a "Secret Steel Breast Plate" or "Secret Armor" (Anon. 1861e:264). Indeed the illustration that accompanied the advertisement for the "Soldiers' Bullet Proof Vest" portrayed a typical military vest, complete with buttons, collar, and watch pockets (Figure 7.4). Vests surviving in museum collections confirm the assessment of one soldier, who described the vests as "nothing more than ordinary vests with metal plates between the lining and outside of the front of the vest" (De Velling 1889:119).

While a few proponents of bulletproof garments overtly confronted the question of cowardice whether by promising that vests would go undetected or by proclaiming the bravery of those men who chose to wear them, other advertisers

The Soldiers'
Bullet Proof Vest

Has been repeatedly and thoroughly tested with Pistol Bullets at 10 paces, Rifle Bullets at 40 rods, by many Army Officers, and is approved and worn by them.

It is simple, light, and is a true economy of life — it will save thousands. It will also double the value and power of the soldier; and every man in an army is entitled to its protection. Nos. 1, 2, and 3 express the sizes of men, and No. 2 fits nearly all.

Price for Privates' Vest, $5. Officers' Vest, $7. They will be sent to any address, wholesale or retail.

Sold by MESSRS. ELLIOTT, No. 231 Broadway, New York, and by all Military Stores. Agents wanted.

Figure 7.4. Advertisement for "The Soldiers' Bullet Proof Vest," *Harper's Weekly,* 15 March 1862. (Photo courtesy American Antiquarian Society, Worcester, Mass.)

approached this conundrum using tendentious, coded language that emphasized the strength, power, value—and, thereby, manliness—a soldier could gain from using a bulletproof vest. Body armor, advertisers claimed, had the ability to confer manliness upon a soldier—"the soldier protected" was "twice the man as the soldier exposed" (Anon. 1862d:3). Furthermore, a soldier who wore a bulletproof vest would "return a wiser and stronger man" (Stokes 1862:5). By this estimation, the garment had the power not only to improve a man's physical strength, posture, and physique but also to strengthen his inner resolve; garments possessed the ability to steel both a soldier's body *and* his mind against attacks from the enemy, thereby *emphasizing*—not undermining—his manliness and grit. "Let the soldier feel that he can defy bullet and bayonet," declared an advertisement in *Frank Leslie's Illustrated Newspaper,* 8 March 1862, "and he will carry everything before him." Owning and wearing a vest would purportedly "double the value and the power of the soldier." On the most basic level, a living soldier was more useful than a dead one. Indeed, the article went on to note that, "nearly every soldier killed loses to the nation his equipment (from $30 to $70) and not only this, but he loses his months of training and his more valuable experience" (Anon. 1862c:252). But this statement about increasing the value and power of the soldier had further implications: with the appropriate battle garments, a Union soldier's bravery would be enhanced, and he could demonstrate his own prowess by achieving an even more glorious victory for the nation. It was precisely in these terms that an incident was relayed concerning the successful use of a bulletproof vest during the Battle of Oak Grove, Virginia. One lieutenant who was wearing a "steel-plated" vest was struck by two bullets in the chest. After which, "he frankly confesses that when he discovered the first ball did not hurt him, he 'was ten times as brave' as he had been" (Moore 1862, vol. 5:232).

Importantly, the material object itself and the technology used to manufacture it conferred this strength and value upon a man by exacting a physical transformation on the body and manipulating his posture. Judgments about a person's character and station in life were made on the basis of posture and the way in which they carried themselves. A tall, erect posture conveyed discipline, self-worth, and manliness. Indeed, as Scott Sandage (2005:56) has suggested, posture was central to representations of success and failure. Mid-century writers and illustrators frequently depicted failed men as stooped beggars, and images of slumped, shuffling slaves reflected developing theories in physical anthropology and phrenology that linked race, body shape, and upright posture to gender and degrees of intelligence (Weicksel 2014). According to advertisers, the bulletproof vest had an advantage over ordinary vests: it would keep "the wearer erect," adding "grace and dignity to his *form*" (Stokes 1863:26). This image of the soldier standing tall, both supported and empowered by his steel vest, directly countered that of the hunched soldier "skulking behind" his armor.

The material components of the vest provided the means of physical, and thereby inner, transformation. Rigid metal plates worn over the breast and secured by hooks or leather straps on the shoulders required a soldier to stand straight, lest the plates press uncomfortably into his abdomen or the straps become dislodged from his shoulders. Other styles of armor, with straps that bound the metal to the man's body, as in the example of the quilted armor, encased the body and had a similar effect on a man's ability to stand straight. Through his erect posture, advertisers claimed, the soldier exuded a new confidence, a confidence that he both conveyed to his comrades and wore into battle.

In addition to asserting control over the movement of the body, the materiality of bulletproof vests of the Smith design was also important to their concealment. The steel breastplates were said to be expertly made such that they could be "so neatly fitted inside of a military vest as not to be noticeable" (Anon. 1861i:2). For those men who had hesitations about the possible connections between cowardice and bulletproof garments, then, the design of the vests was intended to prevent their detection by other soldiers, or so advertisers claimed. In actuality, when considering how the vest as a material object would have fit the body, it is readily apparent that a garment that was available in a maximum of two or three sizes would not have fit the body well. Furthermore, the rigidity of the metal meant that a vest would not move with the body, thereby belying its presence when the body was in motion. And even when a vest was entirely concealed, were a soldier to die, his "secret cuirass" could be found out, tarnishing his bravery and honor in battle and placing his willingness to die for the nation into question. This was not an uncommon occurrence; soldiers wearing armor were found on numerous occasions, having died from wounds to the head and other unprotected areas of the body or because bullets had perforated their armor. After the Battle of South Mountain, for instance, troops of the 30th Ohio walked the battleground: "We found an officer of a South Carolina regiment in a breast plate of hardened steel, thick and heavy, after the fashion of old, fitting close up the neck and under the armpits, and reaching to the waist, with a bullet through his forehead" (quoted in Heineman 2013:160).

TESTING THE VALUE OF BULLETPROOF VESTS

Given that an estimated 94% of Union injuries were caused by bullets, versus 5.5% by artillery fire and less than 0.4% by saber or bayonets, it is clear why efforts to produce protective garments focused on bulletproofing (Faust 2008:41). But how effective was this technology? Soldiers and their families were inundated with positive claims regarding tests performed on body armor in a variety of publications, ranging from advertisements and newspaper articles in the popular press to official-sounding books that included titles such as *The Statistical Pocket Manual of the Army, Navy, and Census of the United States of America*. The majority of published reports were positive. The *Statistical Pocket Manual* (Anon. 1862l:32), for instance, included a blurb that announced:

> An experiment is to be made upon a suit of steel armor, imported as a sample of armor to be worn by officers. It has been subjected to very severe tests, and is capable of resisting even a Minnie ball, at an ordinary distance.

Frank Leslie's Illustrated Newspaper went so far as to include a sketch depicting a vest plate that "received 12 balls with full charge from Tranter's largest size (English) pistol at *ten*, *five* and *two* yards distance." The "value" of the vest, *Leslie's* asserted, was "proved beyond doubt," and two images of flattened rifle and pistol bullets were offered as further evidence of its effectiveness (Anon. 1862c:252). These published testaments to the success of bulletproof technology were, however, often at odds with the opinions expressed in soldiers' private letters. With no official governmental oversight or universal standards for testing, nor the official military issue of body armor, it was entirely the soldier's individual prerogative as to whether or not to accept that the armor had been thoroughly tested.

Many men were unwilling to accept such claims. On a superficial level, soldiers and officers were concerned about themselves or their men being cheated by "Sutlers and other *Sharks*" and wasting their money on worthless goods (Johnson 2004:113). Some officers, driven by "a sense of duty," made a concerted effort to ensure that the soldiers under their command were not taken advantage of by sutlers, peddlers, or advertisers. William Carey Walker (1885:21) recalled a conflict between a colonel and a man who was selling bulletproof vests to soldiers in camp: "Col. Ely, who had often smelt powder in dangerous proximity to bullets, was incredulous of the statement made by the dispenser of the steel vests." Ely took one of the vests, set it up as a target, and shot several holes through it. He then "ordered the arrest of the vender, made him refund to each soldier the amount he had received in exchange for the worthless armor, and gave him the opportunity for reflection in the regimental guard-house." Another unnamed officer purportedly asked the *Hartford Daily Courant* (Anon. 1862a:2) to inform readers that "the officers or privates who invest their six and ten dollars in these articles, will never receive any equivalent for their money."

Men were not merely concerned about money, however; they also expressed an underlying concern about being duped into wearing a superfluous, ineffective garment. If a man were going to don a heavy, cumbersome, hot piece of armor that could also potentially call his bravery into question, it needed to be worthwhile. Conflicting statements about the efficacy of bulletproof armor in the popular press contributed to desires for firsthand knowledge of this technology. Men not only conducted their own tests of these garments but, in at least a few cases, also displayed the results prominently for others to see. In New Haven, for instance, a vest was tested at a distance of 150 yards: "It was well ventilated. A sample of the vest can be seen at Brown & Gross bookstore" (Anon. 1862a:2). Similarly, a vest made of boiler iron and worn by a Confederate captain was displayed at a store in Philadelphia, where all could examine the puncture hole and dents in the metal (Hess 2008:81).

Newspaper articles carried conflicting accounts of the effectiveness of the various styles of armor. Statements about the flaws of armor printed one week were sometimes retracted the next. Shortly after printing a story condemning Atwater's Adjustable Armor (Anon. 1862a), the *Hartford Daily Courant* reported that the writer had "confounded the Armor Vest with Atwater's Adjustable Armor, and we are informed, acknowledges his mistake" (Anon. 1862b:2). The article went on to assure its readers that the Atwater Company claimed their armor would "resist pistol shots at *any distance* and rifle balls at 200 yards" and that it had been "tried at Washington" (Anon. 1862b:2). In a similar retraction, the *Columbian Register* (Anon. 1862i:2) printed: "The armor vests, pronounced by the Norwich regiment to be 'bogus,' were really the well-tried articles made by G. & D. Cook & Co. of this city, and are warranted to be what they are promised." To further allay concerns, harrowing tales of the battlefield in which armor-wearing men were saved from untimely deaths were included in advertisements and journalists' columns. "One man, under McClellan, when before Yorktown, was struck EIGHT times in the BREAST, but not hurt" (Anon. 1862h:3).

Whether or not newspaper editors had an incentive to include such retractions, it is evident that there were discrepancies between advertisers' claims and newspaper reports versus the results of soldiers' own tests. Soldiers reported in letters home that the vests were a failure. In a letter to his wife, Col. Johnson (2004:112) of New Jersey wrote that he and his fellow field officers had just returned from testing "the great, mighty, powerful 'bullit proof' vest and the result is that a common musket put a ball clear through it at 50 yards." He, like Col. Ely, deemed the vests "a humbug! a humbug!!" and "not worth a continental curse!" For many men, such a test resulted in immediate abandonment of the garments. As Sheldon Thorpe (1893:15) recalled, after soldiers' vests had been "rudely perforated with bullets at Camp Chase," they were "ignominiously kicked aside and the skeletons probably repose there to this day." The Smith vest design, in addition to inconsistently resisting bullets, had an inherent flaw. Although the two breastplates were intended to overlap when the vest was buttoned, this seam created a penetrable line along the entire length of the vest, making this armor susceptible to both bullets and bayonets. Such was the case for one man (Haynes 1896:78):

> A rebel had evidently made a desperate lunge at him with a bayo-
> net, the point of which, striking well around to the side, glanced
> along the steel, cutting the cloth in its course, until passing between
> the plates at their junction, it deeply pierced the soldier's breast.

Men did not base their decisions regarding body armor solely on ballistic tests or withstanding bayonet thrusts. They also assessed the level of discomfort the garments inflicted. Even a soldier who accepted the technological claims about bulletproof vests could ultimately reject their use if he found it too painful, uncomfortable, or inconvenient. Wearing steel breastplates introduced another layer of weight to a man's attire. General Robert McAllister (1998:181) wrote: "I did procure an armor vest. But it is too heavy, too warm, and inconvenient to wear in this hot weather. I may put it on some time. I did not have it on at West Point." For some men, the prospect of wearing a vest would have meant a costly change in wardrobe: "I cannot wear a bullet proof vest under my present coat—too small" (quoted in Samito, 1998:71).

Many soldiers laughed at the prospect of encasing oneself in steel, mocking those who might wear the vest by referring to them as "tin stoves" and other derisive terms. One soldier expressed the sentiments of many when he asked,

> What? carry a pound or two of steel around you (in connection
> with all the other paraphanalia that you must carry) with the sun
> pouring down on you, and all merely to protect a few square inches
> of the body, in case a shot might accidentally hit you in that precise
> spot. (Johnson 2004:112)

Lincoln's clerk, William Stoddard (1890:39–40) claimed that the "surliest remark yet made upon" a bulletproof vest was by

> a grim old soldier with stars upon his shoulders: "So that's a cuirass!
> Well the inventor must be a queer ass to think a man could lug that
> thing on a march in a hot sun, or on the double quick."

Many vests were abandoned upon discovering how difficult it was to march carrying both steel plates and a heavy knapsack. Francis Buffum (1882:66) recalled the experiences of the 14th New Hampshire Volunteers on their first day of marching:

> Not a mile of ground had been traversed before a general murmur
> expressed the condition of things. The iron-clad-vested men were
> the first and loudest in their complaints. The lagging, fagged-out,
> green volunteers, panting over a two-mile march, must have been
> objects of ridicule to a veteran of the Army of the Potomac.

Heavy and cumbersome, the armor was at least marginally effective at deflecting shrapnel, ricochets, and shots fired from long distances, as suggested by the dents in the steel of surviving examples of body armor. Furthermore, soldiers' claims that their own or a comrade's life had been saved by wearing armor should not be discounted. Nathaniel Wales, for instance, said that "a bullet struck me just below the heart . . . knocking me down." Wales was not severely wounded, although his vest was dented and his chest was "swollen and for ten days or a fortnight I was unable to draw a long breath" (quoted in Dean 1920:60). One Iowa soldier was wearing a steel vest when he was struck by a bullet, knocked breathless, and reported killed. Imagine the surprise of his comrades when he returned to camp one month later, along with 200 other men who had been taken as prisoners of war (Genoways and Genoways 2001:166).

Years later, during World War I, Bashford Dean, Curator of the Metropolitan Museum of Art's armor collection, proceeded to collect additional test results of Civil War-era vests, coming to the conclusion that they were at least partly effective (Dean 1920:58–60). But the same technology that allowed for this minimal protection also meant that body armor was a dangerous object in and of itself. If a soldier was shot in the chest, the impact itself, more than one soldier claimed, could result in serious injury. As one man quipped, "If the bullet did not go through it would knock a man into the middle of next week so that he might as well be killed first as last" (Boyd 1998:23).

The perforation of the steel plates represented the deadliest scenario. Breastplates—whether or not they were covered in cloth—could inflict fatal harm upon the wearer if they were punctured by a bullet, especially one fired at close range. Colonel Johnson (2004:112) wrote to his wife immediately after he and his fellow officers had completed a test upon a vest in which a common musket put a ball through it—"through yes, and carried some four or five inches of the stuff with it." The metal and cloth fragments carried by the bullet into a man's body, he surmised, "would have killed the devil himself." Such incidents did, in fact, occur, as in the case of Col. William P. Rogers, who was shot at close range while wearing a bulletproof vest. The bullet tore through the metal and into his body, leaving a gaping hole in his chest—a scene preserved by the image captured by a photographer after the vest had been removed. This vest, now preserved in the Wisconsin Historical Society, has a jagged puncture hole two inches in diameter (Hess 2008:82; Zeller 2005:86).

Numerous war memoirs relate incidents in which men, walking the field after a battle for signs of life, discovered dead armor-clad soldiers. Lucius Chittenden (1891:419–420) found a soldier leaning against a tree:

> I placed my hand on his chest to detect any sign of life. It encountered a metal substance. I opened his clothing, and took from beneath it a shield of boiler-iron, moulded to fit the anterior portion of his body, and fastened at the back by straps and buckles.

It was then that Chittenden saw the reason for the soldier's demise: "Directly over his heart, through the shield and through his body, was a hole large enough to permit the escape of a score of human lives." A section of body armor worn in the Battle of Shiloh provides evidence of the devastating effects a bullet could have when it pierced a breastplate. Punctured by two bullets, the jagged metal curls

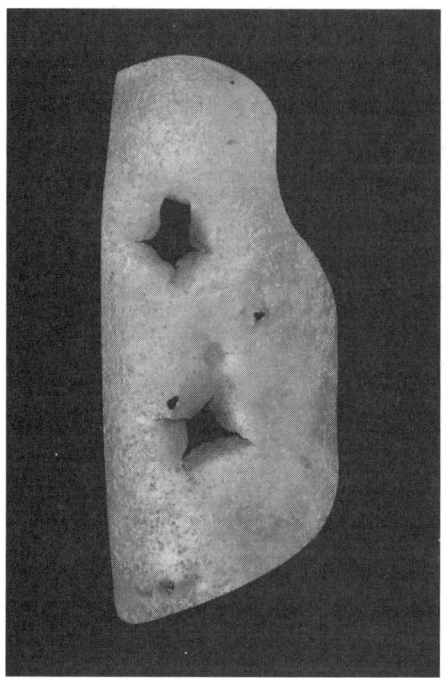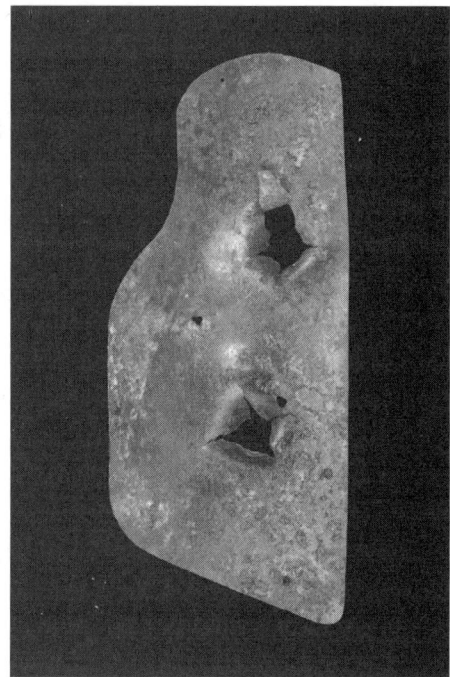

Figure 7.5. Front (left) and back of pierced breastplate found on Shiloh battlefield, 1862. (Photos courtesy Tennessee Virtual Archive, Tennessee State Library and Archives, Nashville, Tenn.)

sharply inward and would have cut the wearer's torso, likely introducing metal fragments into the body (Figure 7.5).

Even if a man were not immediately killed, being hit by a bullet while wearing armor would have increased a man's chances of dying from infection. Such dangerous potential, combined with the number of fatal wounds armor-wearing men suffered in places other than the front torso, suggests that even widespread use of armor vests would likely have had a limited impact on improving the rate of combat casualties. When a man was shot it was essential that a surgeon remove not only the bullet but also the tiny fragments of cloth that were often pushed into the wound on impact. A man wearing the Smith's bulletproof vest introduced extra layers of not only fabric but also metal fragments, in addition to the bullet, that the surgeon would need to locate and remove, a procedure that increased risk of infection. Indeed, although one man was wounded by only a bullet, "the convex side of the dent [made by that bullet] had cut through vest, shirt and undershirt" (Dean 1920:60). Had the bullet entered the body, so too would fragments of steel and fabric. Army surgeons' manuals stressed just how imperative it was to remove even the tiniest bits of cloth, steel, or lead: "When foreign bodies, as balls, pieces of bone, cloth, wadding, etc., are driven into the pleural cavity they produce fatal results, by inflammation and exhausting discharges, unless removed" (Chisholm 1989:300). Technology intended to save a man's life, then, held the potential to inflict even greater harm than a bullet in particular circumstances. The epitaph quoted by Chittenden (1891:419) in relating his discovery of the dead, armor-clad soldier proved quite applicable: "The means employed his life to save, Hurried him headlong to the grave!"

BECOMING TRAGICAL RELICS

By late 1863, while they may have continued to be available for purchase in stores or from peddlers, protective garments seem to have faded from advertisements. Despite a shift away from frontal assaults in massed column in the latter half of the war (Linderman 1987:135–138), the questionable effectiveness and potential danger of wearing body armor outweighed the increasing acceptability of fighting from behind protected entrenchments. Indeed, contrary to *Scientific American*'s positive assessment of this technology, many soldiers did not find a bulletproof vest to be "very strong in proportion to its weight." Hot, heavy, cumbersome objects that were only marginally effective, such protective garments did not fulfill advertisers' promises. Certainly some men were saved by the use of steel breastplates, but others suffered painful deaths when bullets tore through the metal encasing their chests. The rigid steel plates improved the erectness of a man's posture, but their awkward fit slowed soldiers' movement and made marching uncomfortable. While some men felt bolder in battle when protected by steel, most shrunk from its use for fear of being labeled a coward.

The account of Maj. Sidney Willard's death illustrates these conflicted cultural attitudes toward bulletproof vests. According to his eulogist, Cyrus A. Bartol (1862:20), Willard was wearing armor when a Minié ball struck the steel breastplate, "became flattened, glanced off, and entered the groin," causing him to fall off his horse and onto his face. Willard died of his bullet wound "calm and easy" in the field hospital—a sign of his soul's peace. Nevertheless, his eulogist found it necessary to assert that Willard wore the steel vest only because it was given to him "as a token from Col. Wild," an officer whom he greatly respected. Willard, the eulogist continued, "would not have worn [body armor] of his own accord," citing the fact that prior to this skirmish, he had "carried it on his saddle" (Bartol 1862:20). Indeed, by his own admission Willard "did not own or wear a steel vest" prior to receiving Wild's gift (Willard cited in Bartol 1862:37). Willard wore the armor not out of fear, Bartol implied, but rather only out of gratitude for the gift. This miscalculated decision, Bartol asserted, should not overshadow the fact that Willard was "a brave, noble man, and a good commander" who was "well prepared" to "lay down his life for his country!" (Bartol 1862:20). In doing so, the eulogist attempted to offer a resolution to the tension between Willard's use of armor and his brave death.

The introduction of bulletproof garments did not result in a sea change in either cultural understandings of death or military technology. But the brief appearance of bulletproof vests and other forms of body armor reveals important connections among technology, manhood, mortality, and the development of mass culture during the Civil War. Even in the failure of bulletproof technology, the discourses surrounding its manufacture, marketing, use, and abandonment reveal the blurred boundaries and definitions of manhood, duty, and cultural aspects of death. This was a culture astride two ages; a moment in which two competing ideals of manhood and duty coexisted—one that demanded the bold sacrifice of life for the sake of the nation and another, emerging culture in which the protection of life was paramount. Such conflicting ideas of manhood are embodied by the contrasting images of the wartime martyr and the timid, skulking, armor-wearing soldier. Soldiers' debates over, rather than the outright rejection of, the use of body armor suggests that the culture of the Good Death that pervaded the antebellum era was not altogether entrenched and that it was, in fact, beginning to be questioned early in the war and not merely after four years of bloodshed. Men desired to protect themselves yet feared the implications of dying in a "steel cuirass"—to call into question their willingness to die and place into limbo their families' assumptions about the state of their soul. And yet advertisers challenged the idea that patriotic duty required an unflinching willingness to die for one's country. As Alice Fahs (2001:10–11) has argued in the case of popular wartime literature, these advertisers, too, helped to direct and shape readers' and consumers' responses to the war.

By promoting bulletproof garments advertisers created a context in which Americans began to imagine how using technology to preserve one's life could be considered a noble and manly act, one that increased the "value" of the soldier to both his country and his family. To confront the enemy was certainly brave and manly, but to protect oneself from the enemy conveyed an even stronger sense of the meaning of manhood and commitment to both nation and family. Some soldiers, in spite of the derision underlying taunts about wearing "iron stoves" or wearing armor on their backsides, bought into these advertisers' claims in the midst of suffering, death, and loss. And yet, their hesitancy to embrace bulletproof vests and attempts to hide the use of protective garments from comrades reveal cultural tensions over sacrifice. This was a shifting but not yet fully altered culture in which death and manhood were intertwined. The rhetoric armor advertisers injected into the popular press contributed to a discourse in which it became possible to imagine, but not yet fully realize, the compatibility of patriotism and protecting a soldier's mortality by means of new forms of technology.

This disconnect between protecting one's body and retaining one's manhood would eventually be overcome as inventors continued to manipulate combinations of steel, fabric, and ultimately, polymers. Armor would come to play increasingly prominent roles in each future American military conflict. But at a moment in which ballistics technology still far outpaced the repurposed, though repackaged, technology used to manufacture protective garments, the vast majority of men were willing to take their chances on the battlefield. Cumbersome, heavy, and performing poorly in soldiers' own tests, by the end of the Civil War bulletproof vests, steel breastplates, and other styles of personal armor were little more than technological curiosities. Removed from the bodies of dead soldiers, retrieved from shallow, unmarked graves, or unearthed from the sandy banks of Lake Champlain, these garments were preserved as wartime relics—trophies of the battlefield and curiosities of a past time, reminders of the blood that had been shed.

Acknowledgments

I thank Margaret Abruzzo, Kathleen Neils Conzen, Edward S. Cooke Jr., Steve Delisle, Chris Dingwall, Samantha Dorsey, Daniel Greene, Kasey Grier, Jim Grossman, Bart Hacker, Earl Hess, Elisa Jones, Kate Masur, John McCormack, Emily Romeo, Amy Dru Stanley, and Margaret Vining. Special thanks are due to Leora Auslander and Christine Stansell for their suggestions and insight. For research support, I also thank the American Antiquarian Society, Historical Society of Pennsylvania, Library Company of Philadelphia, Newberry Library, and the Department of History, Division of the Social Sciences, and the Center for the Study of Gender and Sexuality at the University of Chicago.

References

Anonymous. 1861a. "The Fight." *Connecticut Courant*, 20 April 1861, 2.

———. 1861b. Steel Clad Steam Chariots of War. *Scientific American*, 5, no. 1(6 Jul.):9.

———. 1861c. National Encouragement to the Novelties of Inventors. *Scientific American*, 5, no. 5 (3 Aug.):73.

———. 1861d. Military Inventions. *Scientific American*, 5, no. 17(26 Oct.):266.

———. 1861e. Secret Steel Breastplate. *Scientific American*, 5, no. 17(26 Oct.):264.

———. 1861f. The Metal for Iron-Clad Ships. *Scientific American*, 5, no. 17(26 Oct.):266.

———. 1861g. "Very Yankeeish." *Fayetteville Observer*, 21 November 1861, 3.

———. 1861h. "State Items." *Hartford Daily Courant*, 21 November 1861, 2.

———. 1861i. "State Items." *Connecticut Courant*, 30 November 1861, 2.

———. 1862a. "Atwater's Adjustable Armor." *Hartford Daily Courant*, 12 February 1862, 2.

———. 1862b. "Atwater Adjustable Armor." *Hartford Daily Courant*, 20 February 1862, 2.

———. 1862c. "The Soldier's Bullet-Proof Vest." *Frank Leslie's Illustrated Newspaper*, 8 March 1862, 252.

———. 1862d. "The Soldier's Bullet-proof Vest." *Philadelphia Inquirer*, 14 March 1862, 3.

———. 1862e. "The Soldiers' Bullet Proof Vest." *Harper's Weekly*, 15 March 1862, 176.

———. 1862f. "Shields for Cowards' Heart." *Daily Picayune*, 3 April 1862, 2.

———. 1862g. "Armor Vests." *Columbian Register*, 10 May 1862, 2.

———. 1862h. "Attention, Soldiers!" *Columbian Register*, 23 August 1862, 3.

———. 1862i. "Local News." *Columbian Register*, 30 August 1862, 2.

———. 1862j. "The 'Soldier's Bullet-Proof Vest." *Boston Daily Advertiser*, 6 September 1862, 2.

———. 1862k. "Bullet-Proof Vests." *New Hampshire Patriot and State Gazette*, 17 September 1862, 2.

———. 1862l. *Statistical Pocket Manual of the Army, Navy, and Census of the United States of America.* Boston: D.P. Butler.

———. 1863. "The Soldiers' Bullet-proof Vest." *Harper's Weekly*, 14 March 1863, 176.

———. 1871. *The Town of Wayland in the Civil War of 1861–1865 as Represented in the Army and Navy of the American Union.* Boston: Rand, Avery and Frye.

———. 1875. "A Mysterious Find." *Columbian Register*, 30 October 1875, 1.

———. 1886. *Biographical and Historical Record of Wayne and Appanoose Counties, Iowa.* Chicago: Inter-State.

Bartol, Cyrus A. 1862. *The Nation's Hour: A Tribute to Major Sidney Willard, Delivered in the West Church.* Boston: Walker, Wise and Company.

Bederman, Gail. 1995. *Manliness and Civilization: A Cultural History of Gender and Race in the United States, 1870–1917.* Chicago: University of Chicago Press.

Billings, John. 1887. *Hardtack and Coffee; or, The Unwritten Story of Army Life.* Boston: George M. Smith.

Boyd, Cyrus F. 1998. *Diary of Cyrus F. Boyd, Fifteenth Iowa Infantry, 1861–1863.* Ed. Mildred Throne. Baton Rouge: Louisiana State University Press.

Bruce, Robert. 2008. *Fighting Techniques of the Napoleonic Age, 1792–1815: Equipment, Combat Skills and Tactics.* London: Amber Books.

Bryson, George G. 1906. Handcuffs on Manassas Battlefield. *Confederate Veteran*, 14, no. 6:304. Nashville, Tenn.: S.A. Cunningham.

Buffum, Francis H. 1882. *The Great Rebellion: Being a History of the Fourteenth Regiment New Hampshire Volunteers.* Boston: Franklin Press.

Cheney, John T. 1862–1901. John T. Cheney Collection. U.S. Army Military Institute Manuscript Collection, Carlisle, Pa.

Chisholm, J. Julian. 1989. *A Manual of Military Surgery: For the Use of the Surgeons in the Confederate States Army*. San Francisco, Calif.: Norman. [Originally published 1861.]

Chittenden, Lucius. 1891. *Recollections of President Lincoln and His Administration*. New York: Harper and Brothers.

Dean, Bashford. 1920. *Helmets and Body Armor in Modern Warfare*. New Haven, Conn.: Yale University Press.

De Velling, Charles. 1889. *History of the Seventeenth Regiment: First Brigade, Third Division, Fourteenth Corps, Army of the Cumberland*. Zanesville, Ohio: E.R. Sullivan.

Dunkelman, Mark H. 2004. *Brothers One and All: Esprit de Corps in a Civil War Regiment*. Baton Rouge: Louisiana State University Press.

Elliott, [Messrs.]. 1862. "The Soldiers' Bullet-Proof Vest." *Evening Post*, 15 April 1862, 3.

Fahs, Alice. 2001. *The Imagined Civil War: Popular Literature of the North and South, 1861–1865*. Chapel Hill: University of North Carolina Press.

Faust, Drew Gilpin. 2008. *This Republic of Suffering: Death and the American Civil War*. New York: Alfred A. Knopf.

Ffoulkes, Charles John. 1912. *European Arms and Armour in the University of Oxford*. Oxford, UK: Clarendon Press.

Foote, Lorien. 2010. *The Gentlemen and the Roughs: Manhood, Honor, and Violence in the Union Army*. New York: New York University Press.

Genoways, Ted, and Hugh H. Genoways, eds. 2001. *A Perfect Picture of Hell: Eyewitness Accounts by Civil War Prisoners from the 12th Iowa*. Iowa City: University of Iowa Press.

Haynes, Martin Alonzo. 1896. *A History of the Second Regiment, New Hampshire Volunteer Infantry, in the War of the Rebellion*. Lakeport, N.H.: printed by author.

Heineman, Kenneth J. 2013. *Civil War Dynasty: The Ewing Family of Ohio*. New York: New York University Press.

Hess, Earl. 2008. *The Rifle Musket in Civil War Combat: Reality and Myth*. Lawrence: University Press of Kansas.

Holley, Alexander Lyman. 1865. *A Treatise on Ordnance and Armor*. New York: D. Van Nostrand.

Johns, Henry T. 1864. *Life with the Forty-Ninth Massachusetts Volunteers*. Pittsfield, Mass.: printed by author.

Johnson, Charles F. 2004. *The Civil War Letters of Charles F. Johnson, Invalid Corps*. Ed. Fred Pelka. Amherst: University of Massachusetts Press.

Linderman, Gerald F. 1987. *Embattled Courage: The Experience of Combat in the American Civil War*. New York: The Free Press.

Malet, William Wyndham. 1863. *An Errand to the South in the Summer of 1862*. London: Richard Bentley.

McAllister, Robert. 1998. *The Civil War Letters of General Robert McAllister*. Ed. James I. Robertson Jr. Baton Rouge: Louisiana State University Press.

McPherson, James. 1997. *For Cause and Comrades: Why Men Fought in the Civil War*. New York: Oxford University Press.

Moore, Frank, ed. 1862. *The Rebellion Record: A Diary of American Events*. 5 vols. New York: G.P. Putnam.

NARA RG 92 [National Archives and Records Administration Record Group 92]. *Records of the Office of the Quartermaster General*. Washington, D.C.: NARA.

Peterson, Harold Leslie. 1973. *Body Armor in the American Civil War*. Chicago: Follet.

Rhodes, Robert Hunt, ed. 1992. *All for the Union: The Civil War Diary and Letters of Elisha Hunt Rhodes*. New York: Vintage Books.

Samito, Christian G. 1998. *Commanding Boston's Irish Ninth: The Civil War Letters of Colonel Patrick R. Guiney, Ninth Massachusetts Volunteer Infantry*. New York: Fordham University Press.

Sandage, Scott. 2005. *Born Losers: A History of Failure in America*. Cambridge, Mass.: Harvard University Press.

Sinkević, Ida, ed. 2006. *Knights in Shining Armor: Myth and Reality, 1450–1650*. Piermont, N.H.: Bunk Hill Publishing, in association with Allentown Art Museum.

Stoddard, William O. 1890. *Inside the White House in War Times*. New York: Charles L. Webster.

Stokes, Charles. 1863. *Charles Stokes and Co.'s Illustrated Almanac of Fashion, 1863*. Philadelphia.

Stokes, Granville. 1862. "Hints to the Drafted." *Philadelphia Inquirer*, 22 October 1862, 5.

Thorpe, Sheldon B. 1893. *The History of the Fifteenth Connecticut Volunteers in the War for the Defense of the Union, 1861–1865*. New Haven, Conn.: Price, Lee and Adkins.

Turner, Terence S. 2007. "The Social Skin." In *Beyond the Body Proper: Reading the Anthropology of Material Life*, ed. Margaret Lock and Judith Farquhar, pp. 83–103. Durham, N.C.: Duke University Press.

Walker, William Carey. 1885. *History of the Eighteenth Regiment, Conn. Volunteers*. Norwich, Conn.: Gordon Wilcox.

Weicksel, Sarah Jones. 2014. Quand l'uniforme fait l'homme libre: Les soldats noirs dans la Guerre civile américaine (1861–1865) [To look like men of war: visual transformation narratives of African American Union soldiers]. *Clio: Femmes, Genre, Histoire*, 40, no. 2:137–152.

Zeller, Bob. 2005. *The Blue and Gray in Black and White: A History of Civil War Photography*. Westport, Conn.: Praeger.

EIGHT

"QUAKER GUN" VERSUS OBSERVATION BALLOON

John A. Macaulay

S pies, cavalry, and telescopes were the traditional intelligence tools available to Civil War strategists, but it was the implementation of the observation balloon that would help change the future of military reconnaissance and weaponry, transforming the war into what Charles Evans (2002) called the "War of the Aeronauts." Indeed, at the beginning of the war, balloon technology, complete with a Union Balloon Corps, made many strategists, including President Abraham Lincoln, believe that observation and intelligence would determine military victory. There was good reason for this. "Sight is, of course, the principal and most immediate medium of real-time intelligence," observed John Keegan (2003:18–20), as long as the commander has "access to a means of communication that considerably outstrips in speed that of the enemy's movement over . . . ground or water." With the Union forces utilizing the telegraph *inside* an aerial Balloon Corps—thereby fusing the media of sight, communication, and speed—strategic, tactical, and intelligence advantages appeared to be with the North.

Yet aside from recognizing that balloons became obvious targets of Confederate munitions, historians have been slow to recognize the flip side of this technology: the tactical ingenuity, imagination, and cost, in the Confederate response to ballooning, and the role this response would have in influencing the Union strategy for victory from that of intelligence to attrition. In the early years of the war Confederate deception went head-to-head with Union ballooning, and both sides expended tremendous effort and costs in these tactical endeavors. In his seminal work *Deception in War*, Jon Latimer (2001:3–5) concluded that "some would say that modern technology renders deception more difficult, but throughout history deceivers have exploited the latest technological developments. . . Deception will no doubt continue just as long as warfare does." And both technology and deception come at a price. Thirty years after the introduction of balloon technology in the Civil War, former Confederate Gen. Porter Alexander assessed the heavy cost of deception. "I have never understood why the enemy abandoned the use of military balloons early in 1863 after using them extensively up to that time," he declared. "Even if the observers never saw anything they would have been worth all they cost for the annoyance and delays they caused us in trying to keep our movements out of sight" (quoted in Rhees 1898:261).

During the Peninsula Campaign, where Union intelligence and reconnaissance relied on balloon technology the most during the war, Confederate Maj. Gen. John Magruder's ingenuous utilization of deception successfully thwarted Union Maj. Gen. George McClellan's advance on Richmond. Latimer (2001:5) concluded that "many of the best practitioners [of deception] have had backgrounds in both the visual and the performing arts," and in the spring of 1862 John Magruder certainly tapped into his thespian past. In grand theatrical style, he used "Quaker guns" (Figure 8.1), cut logs made to look like canons, and dummy encampments, camouflage, ghost fires, exaggerated grunts and noises, and the rotation of troops in a circular fashion in and through the glens of the peninsula, to mask and confuse from aerial view the size and number of Confederate soldiers and artillery. As a result, conflicting reports from the Union balloon *Intrepid* served only to panic, not aid, McClellan (Figures 8.2, 8.3).

These misleading tactics effectively negated the intelligence gathered from the balloon and put a check on the technological advantage the balloon gave to the Union. Instead of clarifying the situation, as intended, the first implementation of aerial observation in the Civil War proved to be no match for traditional tactics of deception. McClellan, informed by balloon reconnaissance, was convinced with an unwavering belief that he was vastly outnumbered and, accordingly, requested reinforcements from Lincoln in order to advance. Meanwhile, though not observed by aerial reconnaissance, Gen. Stonewall Jackson's deceptive diversions in the Shenandoah Valley convinced Lincoln that Washington was vulnerable to Confederate attack, prompting him to halt the deployment of Union Maj. Gen. Irvin McDowell's Corps to the peninsula in response to McClellan's request. Without reinforcements, McClellan delayed the advance on Richmond, giving Confederate Gen. Robert E. Lee time to fortify Richmond's defenses and extend them southward. Whether delivered by Magruder or Jackson, Confederate deception threw a wrench in Union plans at this critical moment and played a direct role in Union failure during the Peninsula Campaign. By creating the

Figure 8.1. Photographer George N. Barnard took this picture of a Quaker gun near Centreville, Virginia, in March 1862. (Photo courtesy Library of Congress, Prints and Photographs Division, Washington, D.C., LC-DIG-cwpb-00942.)

illusion of greater numbers, Confederate deception clouded the Union's strategic vision for many months to come.

Recent scholarship has pointed to the importance of the Peninsula Campaign, this critical moment, and its determinative role in changing Union strategy. Gary Gallagher (2000:3) referred to it as "a Civil War watershed" and concluded that Confederate successes here

> canceled earlier projections of victory within a framework that might
> have restored the Union much as it had existed on the eve of the war.
> Because their premier army had been humbled by the rebels during
> the Richmond campaign, northerners had to confront the prospect
> of pursuing a harsher kind of war to defeat the Confederacy.

Figure 8.2. This May 1862 photo shows Professor Thaddeus S. Lowe at Fair Oaks, Virginia, replenishing the balloon *Intrepid* from the balloon *Constitution*. (Photo courtesy of Library Congress Prints and Photographs Division, Washington, D.C., LC-DIG-cwpb-01563.)

In the months following the campaign, the Balloon Corps, under the civilian supervision of Thaddeus Lowe, would try to keep the unit alive, despite the setbacks of 1862. But in February 1863, Capt. Cyrus Ballou Comstock replaced Lowe as overseer of the corps, signaling that military supervision had displaced civilian. Within a few months, the army (which had never been enthusiastic about the civilian-inspired project) abandoned the corps completely, and Lowe resigned his position. In July, the *New York Times* (Anon. 1863) reported that the Aeronautic Corps of the Army of the Potomac had been "dispensed" with and concluded that "the balloons have been found of no value in the conduct of military operations. This will excite some surprise, for the public had been led to put

Figure 8.3. Professor Thaddeus S. Lowe ascends on balloon *Intrepid* to make observations near Fair Oaks on 31 May 1862. (Photo courtesy Library of Congress Prints and Photographs Division, Washington, D.C., LC-DIG-cwpb-01560.)

considerable confidence in balloon reconnaissance from facts heretofore given." Edwin Fishel (1998:5) concluded that the balloons were "handicapped by fog, wind, and terrain that hid their targets, [and] bulky gas-generating apparatus prevented them from accompanying the army except in the more static situations." But there was more to this failure than army disdain and logistical problems associated with the balloons. The decline of Union aerial reconnaissance coincided with the success of Confederate military deception. As the North came to accept this handicap, Union strategists shifted from intelligence and "confidence in balloon reconnaissance" toward a "harsher kind of war."

In the early months of the war, Lincoln himself had been a subscriber to that confidence. Within days of Union defeat at Bull Run, Lincoln called upon Lowe to demonstrate the advantages of balloon reconnaissance. With a telegrapher in tow, on 16 June 1861 Lowe ascended in the balloon *Enterprise* from a lot near the White House to an elevation of more than 1,000 feet. He marked the event by successfully sending telegrams to both the president and the War Department (Lowe 1861). Impressed by this, Lincoln invited Lowe to spend the night and confer with him about its possibilities. According to Lowe (2004:74), Lincoln "expressed the thought that had General McDowell had the information that only observations from a balloon could give, the result [at Bull Run] might have been different." Lincoln then penned a note of introduction and endorsement to Gen. Winfield Scott on 25 July 1861 (Lincoln 1861). When Scott failed to take the bait and refused to meet with Lowe, Lincoln accompanied him personally to the War Department and declared,

> General, this is my friend Professor Lowe, who is organizing an Aeronautics Corps for the Army, and is to be its Chief. I wish you would facilitate his work in every way . . . and give him all necessary things to equip his branch of the service on land and water. (Lincoln 1861)

Scott reluctantly obliged.

Like Lincoln, McClellan was also a fan of the balloon. Evans (2002:106) concluded that McClellan "proved to be a major ally to the Balloon Corps. . . . Always eager to seize upon new military technology, McClellan used the Corps during a number of major campaigns." This affinity for the corps and new technology makes McClellan's conviction of Confederate numerical strength on the peninsula that much more intriguing. And yet, the question of why he held to this conviction remains largely unanswered. Stephen W. Sears (1992:346) concluded that McClellan "would insist to everyone—the authorities in Washington, his political allies at home, newspapers friendly to his cause—that he had been attacked in the Seven Days by a Confederate army of 200,000." That this conviction gained traction beyond military circles, escaping War Department correspondence, spilling out into civilian reports, gaining wide acceptance among Democrats and even notable Republicans, makes the question even more poignant. Michael C. C. Adams (1978:99) has argued that McClellan's failure to defeat a smaller rebel army created a crisis of confidence in the North and promoted a widespread belief that Confederate soldiers and generals were most likely inherently superior to Union counterparts. With such a belief, one can understand why McClellan's own troops would echo the war call for "reinforcements" and why many Northerners would repeat the cry as often as they did and blame the War Department for Union failure during the Peninsula Campaign. These calls for reinforcements added credence to McClellan's conviction of "vastly superior numbers." George Templeton Strong (1962, vol. 3:236–237, 239), a staunch Republican, confided in his journal that McClellan had been "beat back by a superior force but not destroyed." Siding with McClellan, Strong concluded that "the enemy was superior because we have been outgeneraled. The blame rests, probably, on the War Department. The remedy is speedy reinforcement."

But what reconnaissance did McClellan base his understanding of Confederate numbers on? What information did aerial reconnaissance gather and report to McClellan and when, and did this information run counter to McClellan's conviction of Confederate superiority? On the ground, McClellan availed himself of America's top detective, Allan Pinkerton, for intelligence. Pinkerton served as head of the Union Intelligence Service in 1861–1862. Ahead of the Union invasion of the peninsula, Pinkerton sent ground scouts, often working "undercover" and behind enemy lines, to ascertain the strength of Confederate forces. Pinkerton's reports to McClellan reveal a graduated escalation of Confederate numbers from late March to late summer of 1862. McClellan's official 29 March report, at the "moment of invasion" on the peninsula and based on Pinkerton's analysis from an October 1861 survey, estimated that the "total rebel forces on the York and James River Peninsula . . . [was] 25,000 men." Brigadier Gen. Fitz John Porter tapped the enemy's advanced line at Fort Monroe at 10,000 fewer (15,000), and on 3 April McClellan (1989:227), reflecting that number, telegraphed Secretary of War Edwin M. Stanton that he expected to face an entrenched "force of some fifteen thousand of the rebels." A month later, on 3 May, Pinkerton's number spiked remarkably, escalating in range from "100,000 to 120,000." Seven weeks later on 26 June, McClellan reported 180,000 men, and six weeks later, on 14 August, that number had swelled to 200,000 (Allan 1985). As Sears (1992) pointed out, McClellan's 15,000 estimate in his 3 April communique was notable for two reasons: first, its accuracy, with an overstatement of only 1,400 men, and second, its timing. "It proved to be the only time during the entire Peninsula Campaign," Sears (1992:30) concluded, "that Federal intelligence came anywhere close to an accurate count of the opposing army." So what accounts for the discrepancy, and what, if anything, did McClellan do to confirm that later figure?

Charles Evans (2002) mistakenly assigned the 3 May post-invasion 100,000 figure to McClellan's 29 March pre-invasion intelligence. Evans estimated 100,000 or more Confederates "occupied the area across the peninsula between Yorktown and the Warwick River," sealing off the route to Richmond (Evans 2002:179). He concluded that had McClellan decided to dispatch the Balloon Corps to scout the area, an aerial observation would probably have revealed [Magruder's] deception. But as it remained, McClellan was thoroughly convinced that his army was looking at the prospect of a bloody and prolonged siege.

It is not clear if McClellan's telegram to Secretary of War Stanton preceded or followed his order to Professor Lowe, but McClellan did send an order to Lowe on 3 April requesting that he "accompany General Porter in his advance to Yorktown" (McClelland 1862). Moreover, McClellan, at this moment at least, wasn't as committed to a siege as Evans would have us believe. The same day as his messages to Stanton and Lowe, McClellan (1989:225) wrote his wife, Mary Ellen, and communicated his desire for a quick and rapid move to Richmond.

> I hope to get possession of Yorktown day after tomorrow. Shall then arrange to make the York River my line of supplies. The éclat of taking Yorktown will cover a delay of the few days necessary to get everything in hand & ready for action. The great battle will be (I think) near Richmond as I have always hoped & thought. I see my way very clearly—& with my trains once ready will move rapidly.

But this optimism quickly faded as McClellan learned that the roads that had previously been reported to him as "good" and "natural" with sandy surfaces and good drainage were, in fact, poor, muddy, and practically impassable. Heavy and relenting rains had made them literal mud pits that, as one New Yorker described, came up to "our knees" (Laughton 1862).

At this juncture McClellan's ground intelligence clearly needed aerial reinforcement as both Confederate numbers and Virginia roads appeared to be "muddier" than he initially thought. Given his insistence upon the accompaniment of the Balloon Corps to his Peninsula Campaign, it is clear that McClellan's support of aerial reconnaissance was unmistakable. What is not as clear is which approach, ground or aerial, McClellan relied upon more at any given moment. By all indication, McClellan utilized and reported ground data more than aerial, especially in reporting Confederate numbers. After all, McClellan cites Pinkerton's numbers much more frequently than Lowe's. This is understandable given what McClellan knew about Lowe's record. During July and August 1861, as Confederates retreated to Munson's Hill after First Manassas, Lowe had reported that Confederate forces and numbers were strong. But after Confederates retreated in September and Union forces discovered the deception, Lowe quietly realized how inaccurate his observations of the hill had been. And even though he did not accept responsibility for the lapse in intelligence, he did take credit for naming the "stove pipes" and cut logs that Union forces discovered as "Quaker Guns" (Robinson 1986:12).

From 4 to 12 April 1862, when Confederate Gen. Joe Johnston was placed in command, Confederate Gen. John Magruder stood alone between Fort Monroe and McClellan's advance on Richmond. With the daunting recognition that he was vastly outnumbered without reinforcements of men or artillery, Magruder faced the enemy with a dogged determinism and a deceptively resourceful plan. On 5 April when he saw Union troops for the first time, Magruder ordered his men out of their entrenchments and sent them through enemy fire on a curious expedition. As Capt. James H. McMath (1862) of the 11th Alabama described it, we marched "until we got out of sight just around the point of a hill. We were halted there some ½ hour, when we were counter-marched over to the place we started from." To lend credence to the effect, Magruder ordered his men, once behind the cover of woods, to yell, grunt, drum cadence, blow bugles, and fire periodically. That night, Edmund D. Patterson (2004:17) from Alabama took a few minutes at the end of a very long day to record these events:

> This morning we were called out by the "Long roll" and have been traveling most of the day, seeming with no other view than to show ourselves to the enemy at as many different points of the line as possible.

The next day, 6 April, faced with what appeared to be conflicting reports of Confederate strength, McClellan ordered Lowe to make an ascension as soon as possible and specifically asked the professor to "look for the movement of wagons and teams; and also where the largest number of men are" (Locke 1862). Lowe (1863:273) obliged and made his report to Gen. Porter. From all accounts, it appeared that Lowe's suspicions about Confederate strength and Magruder's antics were raised based on his subsequent request of Porter and others. Lowe asked the general to ascend with him so that "he might judge for himself of the number of the enemy and strength of their works" (Allan 1985:273–274). Porter did so, and they ascended to an elevation of 1,000 feet and stayed there almost two hours in observation. Later that afternoon, Lowe sent up several draftsmen to sketch maps and enemy positions, and Porter ascended once again with Count de Paris. At sunset Gen. Butterfield went up as well. With the frequency of ascensions and the numbers and ranks of observers, aerial observation and assessment of Confederate numbers were strong and objective. Lowe (1985:274) concluded that "the observations and maps thus made were of the greatest importance, and readily enabled the commanding officer to decide what course he would pursue."

Given the events of the day Lowe's statement is very curious and warrants careful scrutiny. Evans (2002:180) saw significance in Lowe's assessment that the "observations and maps" were of "the greatest importance." On the surface, it appeared that Lowe's report to McClellan suggested that he and others were clued into Magruder's tactics and were aware, through balloon reconnaissance, that enemy numbers were not as great as the Confederates tried desperately to portray them. Evans concluded that in spite of this information, "the reality was that McClellan remained pat. Though aerial observation revealed that Magruder's troop strength was not as formidable as first thought, McClellan opted to remain cautious with his approach toward the capture of Yorktown" (Evans 2002:180).

But at this point McClellan's assessment of Confederate strength was still around 15,000, not the 100,000 that Evans has assigned to him. Moreover, Lowe failed to record or file in official records the maps or observations he mentioned. However, he did offer after a month of steady reconnaissance a regret for this lapse: "I regret that I have not more copies of reports, but as I had my camp at headquarters I usually made my reports verbally, assisted in my explanations by references to maps" (Lowe 1863:275). More importantly, his statement that these items "enabled the commanding officer to decide what course he would pursue," (Lowe 1863:274) stands in direct contradiction of the course one would expect McClellan to take if he were informed of Magruder's antics, as Evans would have us believe he was. If McClellan was aware of these theatrics, one would expect him to "move rapidly" as he told his wife he would. Instead, he stalled.

Meanwhile, Magruder realized that, at best, he could only buy time, not stall or stop McClellan's advance permanently. He had anticipated this and had already let Lee know what his next move would be. Three weeks earlier, on 14 March, Magruder (1862:66–67) warned, "Should the enemy advance I should be compelled to withdraw at once the few troops I have in front to my mai[n] line of defense behind Warwick River." True to his word, as McClellan slowly moved up the peninsula, Magruder retreated behind the Warwick and built a defensive line as best he could. With a 13-mile line to defend and only 15 guns, Magruder made up the defense with Quaker guns, hoping to later replace them with real ones. As McClellan approached, Magruder mixed the Quaker guns with real ones along the line to create the illusion of a strong defensive line. By all accounts, Magruder hit pay dirt, and the Confederates simply repeated the same deceptive scenario that they had at Munson Hill, even under the now watchful and informed eye of Lowe's aerial reconnaissance. As Confederates quietly pulled out of Yorktown under the cover of night a few weeks later, Lowe seemed as surprised as anyone.

Later, Brig. Gen. Erasmus D. Keyes requested that Lowe give a statement of his observations of 3 May in the area near the Warwick courthouse. In that statement, Lowe (1863:277) made a lone reference to Confederates' "apparent" installation of earthworks around the courthouse, but it did not alter his belief that they were entrenched with no indication of retreat. He clarified his conviction by explaining that "it was known by all who had an opportunity of knowing that the enemy . . . kept up appearances until the night of the evacuation." Seemingly aware that this explanation was insufficient, Lowe continued in an attempt to justify why the balloons did not detect enemy movement before the night in question. "It is true," Lowe acknowledged, that "army wagons were daily seen plying between Yorktown and Williamsburg, and so reported, but it was impossible to say which way they were loaded."

With this kind of uncertainty coming from aerial intelligence, Union strategists faced the increasing likelihood that the Balloon Corps had become a military liability at the same moment that Confederate strategists, encouraged by the success of what Lowe called "keeping up appearances," celebrated the asset that deception had become. Magruder, a few weeks later, divided Lt. Robert Miller's 14th Louisiana regiment into two parts and kept them marching for 24 hours, or until "reinforcements came." By the time Miller stopped, his regiment had marched from Yorktown to the James and back six times, all the while cementing McClellan's conviction that rebel numbers were large and superior (Casorph 1996:232). Considering the escalation of McClellan's "superior numbers" claim and the national traction it garnered during the Peninsula Campaign, one would expect Lowe to enter into the debate to correct the record. After all, the very life, integrity, and legitimacy of his Balloon Corps were at stake. He did not. Lowe had come to his own defense when he failed to recognize that the Confederates had evacuated Yorktown, but on the question of Confederate numbers, Lowe remained silent.

Jon Latimer (2001:3) indicated that "successful deception is an art rather than a science, although science increasingly provides the technical means by which deception is created." Faced with a Union Balloon Corps watching his every step and movement, Magruder knew this, turned science on its head, and put a check on the technological advantage the balloon, theoretically, had given to the Union. Instead of offering real-time intelligence as intended, the first implementation of aerial observation in the Civil War proved to be no match to the traditional art of deception. In a war astride two ages, art trumped science in the spring of 1862 and, for that moment at least, gave the tactical advantage to the Confederacy.

References

Adams, Michael C.C. 1978. *Our Masters the Rebels: A Speculation on Union Military Failure in the East, 1861–1865.* Cambridge, Mass.: Harvard University Press.

Allan, Maj. E.J. [pseudonym of Allan J. Pinkerton]. (1880) 1985. Reports to Maj. General George C. McClellan, submitted to Hdqrs. Provost-Marshal-General, Army of the Potomac, 3 May, 26 June, 14 August 1862. In *War of the Rebellion: Official Records of the Union and Confederate Armies,* ser. 1, vol. 11, pp. 264–272. Reprint, Harrisburg, Pa: The National Historical Society. [Originally published in War Department et al. (1880–1901).]

Anonymous. 1863. "Use of Balloons in the War." *New York Times,* 12 July 1863, 8.

Casorph, Paul D. 1996. *Prince John Magruder: His Life and Campaigns.* Hoboken, N.J.: Wiley.

Evans, Charles M. 2002. *War of the Aeronauts: A History of Ballooning in the Civil War.* Mechanicsburg, Pa.: Stackpole.

Fishel, Edwin C. 1998. *The Secret War for the Union: The Untold Story of Military Intelligence in the Civil War.* Boston: Houghton Mifflin.

Gallagher, Gary W. 2000. "A Civil War Watershed: The 1862 Richmond Campaign in Perspective." In *The Richmond Campaign of 1862: The Peninsula & the Seven Days,* ed. Gary W. Gallagher, pp. 3–27. Chapel Hill: University of North Carolina Press.

Keegan, John. 2003. *Intelligence in War: Knowledge of the Enemy from Napoleon to Al-Qaeda.* New York: Alfred A. Knopf.

Latimer, Jon. 2001. *Deception in War: The Art of the Bluff, the Value of Deceit, and the Most Thrilling Episodes of Cunning in Military History, from the Trojan Horse to the Gulf War.* Woodstock, N.Y.: Overlook Press.

Laughton, Joseph B. 1862. Letter to family, 14 April. Joseph B. Laughton Papers, Perkins Library, Duke University, Durham, N.C.

Lincoln, Abraham. 1861. Letter to Gen. Winfield Scott, 25 July. Thaddeus S.C. Lowe Collection, National Air and Space Museum, Washington, D.C.

Locke, Fred T. 1862. Message to Thaddeus S.C. Lowe, 6 April. In War Department et al. 1880–1901, ser. 3, vol. 3, p. 273.

Lowe, Thaddeus S.C. 1861. Letter to Abraham Lincoln, 16 June. The Abraham Lincoln Papers. Series 1. General Correspondence 1833–1916. Library of Congress, Washington, D.C.

———. 1863. Report [of operations in the department of aeronautics] to Secretary of War Edwin M. Stanton, 4 June. In War Department et al. 1880–1901, ser. 3, vol. 3, pp. 252–319.

———. 2004. *Memoirs of Thaddeus S.C. Lowe, Chief of the Aeronautic Corps of the Army of the United States during the Civil War: My Balloons in Peace and War.* Ed. Michael Jaeger and Carol Lauritzen. Lewiston, N.Y.: Edwin Mellen Press.

McClellan, George B. 1862. Message to Thaddeus S.C. Lowe, 3 April. In War Department et al. 1880–1901, ser. 3, vol. 3, p. 273.

———. 1989. *The Civil War Papers of George B. McClellan.* Ed. Stephen W. Sears. New York: Ticknor and Fields.

Magruder, J. Bankhead. 1862. Report to Robert E. Lee. In War Department et al. 1880–1901, ser. 1, vol. 9, pp. 65–67.

McMath, James H. 1862. Diary, 5 April. Alabama Department of Archives and History, Montgomery.

Patterson, Edmund Dewitt. 2004. *Yankee Rebel: The Civil War Journal of Edmund Dewitt Patterson*. Ed. John G. Barrett. Knoxville, Tenn.: University of Tennessee Press.

Rhees, William Jones. 1898. Reminiscences of Ballooning in the Civil War. *Chautauquan*, (June):261.

Robinson, June. 1986. The United States Balloon Corps in Action in Northern Virginia during the Civil War. *Arlington Historical Magazine*, 8, no. 2:5–17.

Sears, Stephen W. 1992. *To the Gates to Richmond: The Peninsula Campaign*. New York: Ticknor and Fields.

Strong, George Templeton. 1962. *The Diary of George Templeton Strong*. Ed. Allan Nevins and Milton Halsey Thomas. 4 vols. New York: Macmillan.

War Department et al. 1880–1901. *War of the Rebellion: Official Records of the Union and Confederate Armies*. 128 vols. Washington, D.C.: Government Printing Office. [Reprinted Harrisburg, Pa.: The National Historical Society, 1985.]

NINE

DREAMS OF AERIAL
NAVIGATION

Tom D. Crouch

"Without Eccentricity, there is no Progression."
Solomon Andrews, 1866

The soldiers of the Army of Northern Virginia had little enough to celebrate during the holiday season of 1864. Atlanta had fallen on 1 September, setting the stage for President Lincoln's re-election in November and the end of southern hopes for a negotiated settlement with a Peace Democrat. "Uncle Billy" Sherman had then cut himself loose from his supply lines and marched to the sea, concluding his campaign with the capture of Savannah on 21 December. George B. Thomas had thrown John Bell Hood back at Nashville on 16 December. Their own army was pinned in a growing set of entrenchments at Petersburg, the last bastion standing between the Army of the Potomac and the capture of Richmond.

David E. Johnston, then an 18-year-old private in the 7th Virginia, recalled many years later that, at this dark moment in this history of the Confederacy, "there appear[ed] on our lines a man . . . who proposed . . . to . . . build a machine to navigate the air, carry shells and drop them on the Northern armies, and their cities" (Johnston 1980:273). A comrade, Pvt. Franklin Lafayette Riley (1988:228), confirmed that their unit had attended a lecture given by Richard Oglesby "Bird" Davidson, who was circulating through the lines collecting the money with which to build his flying machine—one dollar each from the enlisted men and five dollars from the officers. "In 6 days, he says, the Yankees could be defeated. On the 7th day, they would surrender." The men of the 7th Virginia concluded that Davidson was a crank and refused to contribute. Other units proved more generous and enthusiastic. One man from Benning's Brigade of Georgia infantry recalled: "I was very anxious to see that man stampede the Yankees" (as quoted in Hess 2009:242). Even the skeptical Pvt. Johnston concluded, "This fellow was only a little ahead of this time."

The Petersburg Campaign had begun in June 1864. The Army of the Potomac faced Robert E. Lee's Army of Northern Virginia at Cold Harbor, 10 miles northeast of Richmond. Union Gen. Benjamin Butler, bottled up in the Bermuda Hundred east of the Confederate capital, noted that southern troops were being pulled from positions on his front, near Petersburg, a rail center that was the key to Richmond. On 9 June 1864, his Army of the James attacked the trenches 10 miles east of the city. While Butler's attack was repulsed, Gen. Ulysses S. Grant was able to slip across the James River and renew the attack, which led directly to a siege of Petersburg.

Throughout this period, the pressing need for reconnaissance plagued Butler. Colonel Edward Wellman Serrell, Chief Engineer of the Army of the James, proposed a revolutionary solution. For some years he had been fascinated by the possibility of a rotary wing craft and had even prepared rough drawings of the *Reconnoiterer*, the aerial craft he envisioned. In addition, he constructed a small hand-launched helicopter model with metal rotors to demonstrate the basic principle. A pull of a string wound around the shaft of the toy, and it rose 100 feet into the air. Impressed, Butler approved leave for Serrell, who would travel to Philadelphia to raise the funds with which to build a larger experimental version of the model (Anonymous 1864c:135).

While Davidson was circulating through Confederate lines and Col. Serrell, with money in hand, was beginning to cut metal in New York, Solomon Andrews, the Mayor of Perth Amboy, New Jersey, was preparing to demonstrate a flying model of his "Aereon" in the halls of Congress and the library of the Smithsonian Institution. Having witnessed the test, Joseph Henry, Secretary of the Smithsonian and the unofficial chief scientist of the United States, reported to Secretary of War Edwin M. Stanton that it was possible "that he can really perform what he has asserted he can do." That being the case, the committee recommended that the War Department fund the construction of a full-scale technology demonstrator (Bache et al. 1864).

In 1840, Alexis de Tocqueville (2003:410) famously suggested that the Americans were a fundamentally practical people who wasted little time on "theoretical science." As the critic and historian Perry Miller (1979) noted, however, Tocqueville was ignoring American confidence in the power of technology to shape a bright future in which wonders once thought to be impossible would be achieved.

However perceptive in other areas, the French observer simply could not comprehend the passion with which these people "flung themselves into the technological torrent, how they shouted with glee in the midst of the cataract, and cried to each other as they went headlong down the chute that here was their destiny, here was the tide that would sweep them toward unending vistas of prosperity" (Miller 1979:197). Support for Miller's observation is to be found in the surprising level of enthusiasm for the possibility for heavier-than-air flight to be found in antebellum and Civil War America. While many, probably most, Americans regarded the possibility of heavier-than-air flight as being something akin to perpetual motion, a substantial number of would-be aviators announced a broad range of plans to take to the sky in winged craft.

Much scholarly attention has been given to the impact of technology on the American Civil War. In addition to the extent to which the railroad, the telegraph, and modern industrial production shaped the conflict, revolutionary weapons such as the submarine pushed contemporary technology to the very limit—and beyond. The airplane, however, represented a much more difficult technical challenge. Yet a significant number of would-be inventors took up that challenge with a view to the destruction of the enemy. What impelled these men and others to believe that they could achieve a technological goal that most men and women still regarded as the very definition of the impossible? What was the source of their technical ideas? What was the reaction to their plans for the conquest of the air? Any attempt to answer those questions must begin by considering the historical context of aeronautical thought as it had evolved during the first half of the nineteenth century.

EARLY AMERICAN FLYING MACHINE SCHEMES

The earliest public description of an American flying machine project came on 22 March 1822, when Rep. William Milnor of Pennsylvania introduced a petition on behalf of one James Bennett, a Philadelphia mathematician, who claimed to have invented "a machine by which a man can fly . . . through the earth's air, can soar to any height, steer in any direction, start from any place and alight without risk of injury" (Anon. 1822a:79). The petition sought to give Bennett a 40-year monopoly on the operation of a flying machine "through that portion of the earth's air which passes over the United States, or so far as their jurisdiction may extend." Less than two weeks later, on 1 April, Rep. Elias Keyes of Vermont introduced a second petition alleging that David Lee, another Philadelphian, was the real inventor of the flying machine in question (Anon. 1822b:95). Both petitions were submitted to a congressional committee from which they did not emerge. When David Lee asked the aging Thomas Jefferson for his opinion, the elder statesman replied that he did not see a future for flight "by mechanical means alone in a medium so rare." He closed his reply "with more good will than confidence" (Jefferson, 1822).

Not all Americans would have agreed with the Sage of Monticello. The following year James Buchanan of Hopkinsville, Kentucky, announced the development of a steam power plant to propel what one newspaper referred to as "the Kentucky Eagle." The engine was "composed of one capillary tube or more, arranged as a steam generator . . . 200 times more powerful than similar devices" (Anon. 1823:2). The flying machine that it was to power would enable

> the citizens of Washington [to] attend dinner parties in Boston and
> return home the same evening; the mail can be carried from the
> seat of Government to the most distant point of the Union; and
> our merchants may visit Europe, transact their business, and be
> home in a week (Anon. 1823:2).

Early the next year, the London *Mechanic's Magazine* (Anon. 1824:222) announced that a gentleman from Lexington, Kentucky, had tried to ascend in what we can presume was a heavier-than-air machine, "which obstinately refused to give up its hold on the earth!" It seems fair to assume that this was James Buchanan, who, having tried and failed, was quick to assure spectators that he was not discouraged.

By 1834 the number of flying machine experimenters and builders for whom we have more detailed information was increasing. "Perhaps it is not generally known," reported the *Liberty Hall and Cincinnati Gazette* on 24 June (Anon. 1834b), "but one of our ingenious local citizens has invented, and has now in preparation, the model of an *aerial steamboat*, in which he proposes to ascend on the fourth of July." While the reporter had "but little expectation of the success of the experiment," the inventor was said to be "very sanguine, having already made (to him) a very successful experiment." The hull of the craft was shaped like a boat, the ribs of which were covered in silk, "to render it very light." A two-horsepower steam engine positioned in the center of the boat, turned "four vertical shafts projecting over the bow and stern into each of which are fixed four spiral silken wings which are made to revolve with sufficient velocity to cause the vessel to rise." A "moveable silken cover designed to assist in counteracting the gravitating force, [while] at the same time tending to assist in its propulsion forward" was also included. The craft weighed only 60 pounds and had cost $300 to build. Until the Fourth of July, the craft was on display "on Race Street nearly opposite the old Lathe factory, below Third St." (See also Anon. 1834c:502.)

The Fourth of July in 1834 came and went without a flight. On 23 August, the *Daily Cincinnati Republican* reported that the "Aerial Steamboat" would be displayed at the Commercial Exchange early the following week. "Mr. Masson, an ingenious mechanic, has spent some months constructing this vehicle, in which he expects to navigate the air by the force of steam." "There is", the paper noted, "nothing of the balloon principle connected to this apparatus," which was to be "elevated and propelled by machinery in the shape of wings." Having inspected the craft, the reporter was unwilling to predict success or failure but assured readers that it was "a beautiful and ingenious piece of mechanism" (Anon. 1834d).

On 22 October 1834 the inventor, signing himself A. Masson, announced in the pages of the *Daily Cincinnati Republican* that he had opened a subscription, hoping to raise $1500 to cover his expenses (Anon. 1834e). He was quick to assure potential supporters, however, that no money need be paid until he had demonstrated the capacity of his machine to fly. That would seem to remove the inventor from the ranks of mountebanks and confidence men. Six days later, another article, this in the *Cincinnati Chronicle and Literary Gazette* (28 October 1834), explained that the four vertical spindles described in the earlier account were driven by leather bands powered by the engine.

> Upon each of these spindles are placed four wings, shaped like a paper fan, when open, with the broad end from the spindle; these wings are not horizontal, but one edge is raised higher than the other. When the spindles are made to revolve, the wings, thus inclined, strike the wind with so much of their broadside as to occasion considerable resistance and the consequent tendency is to make each wing, instead of round against this resistance, to move at an angle upward, cutting the air with its edge. It is in a manner screwing up into the air. (Anon. 1834f)

At this point A. Masson and his Aerial Steamboat vanish from the pages of the city's newspapers. Who was this fellow, and what can be made of his valiant attempt to build and fly a heavier-than-air craft? *The Cincinnati Directory* for 1834, the year in which the craft was built and exhibited, lists an Albert Mason, a steamboat mate living on East Front Street (Anon. 1834a). Given the nautical references in all of the accounts, and the fact that a steamboat mate would be familiar with the propulsion system described, it does not seem too much of a stretch to regard this fellow as a prime suspect for our "ingenious mechanic."

John Pennington (1838), a Baltimore native, began his aeronautical career in 1838 when he published plans for his "steam kite, or Inclined Plane for Navigating the Air." The central feature of the craft was to be a large overhead elliptical plate that generated lift. The fuselage of the craft was shaped like the stock and barrel of a very long rifle. Twin paddle wheels mounted in the wing were to be powered by an internal combustion engine, fueled by a mixture of turpentine and alcohol, delivering one-and-a-half to three horsepower. That pioneering internal combustion engine had been developed by steamboat pioneer Capt. Samuel Moray of Fairlee, Vermont, who was said to have developed this lightweight power plant with a flying machine in mind. Pennington would eventually abandon heavier-than-air flight in favor of a powered airship.

CIVIL WAR FLYING MACHINES

Some notion of interest in flying machines on the part of technically minded Americans is to be found in the fact that between 1842, when John Pennington received a patent for his steam kite, and the end of the Civil War, 18 patents were issued for heavier-than-air craft or navigable airships. Perhaps a better measure of interest in aeronautics is to be found in the fact that from its founding in 1845 through 1865, *Scientific American*, whose founder, Rufus Porter, was a leading proponent of the steam-powered airship, published a total of 131 articles on aeronautics, 57 of which dealt with aerial navigation in one form or another. Clearly the flying machine was a subject of growing interest to Americans interested in technology (Anon. 1860c:165).

Thaddeus Hyatt, a wealthy New York inventor said to hold a patent worth "many thousands of dollars a year," created a stir in the pages of *Scientific American* in August 1860 when he offered a $1000 prize to anyone who could produce a practical flying machine by 1 September 1861 (Anon. 1860a:88). "Whatever advantages the inhabitants of the air may possess by nature," Hyatt remarked, "can be more than matched by . . . ingenuity and skill." One of the early respondents (Anon. 1860b:116) applauded Hyatt's approach but pointed out that even one million dollars was insufficient reward for so great an achievement. "No one invention of man has ever yet made such a revolution in human affairs as a . . . true . . . flying machine would make."

The editor of the magazine underscored that point in early September, arguing that of all imaginable inventions "there is none which so captivates the imagination as that of a flying machine" (Porter 1860:125). Offering general thoughts on aircraft propulsion, the article was illustrated with images of a fellow hanging onto a rocket for dear life, a helicopter with rotors propelled by hot gases exiting from the blade tips, and a "bird woman" carried aloft on a giant kite or wing. The editor also included one interesting "hint to inventors who desire to enter this enticing field. . . . The newly discovered metal aluminum," he advised, "from its extraordinary combination of light weight and strength is the proper material for flying machines."

In the months to come the editors would continue to publish occasional aeronautical news from Europe and general thoughts on technology useful to flying machine experimenters (Anon. 1864a, d). The columns would contain letters from enthusiasts, like the fellow who insisted that the answer was to "discard all inflation" and to pay attention to nature, and to the design of kites, in order to craft a useful winged craft (Anon. 1864b). There were also occasional dissenting voices. "Pneumatics" presented a series of arguments against the possibility of heavier-than-air flight, so that "those who still have any faith in its accomplishment . . . may see the fallacy of all such attempts and turn their ideas into a more profitable channel" (Anon. 1864b).

Over the four years of war, a number of specific flying machine projects were recorded in the pages of the *Scientific American* and in the records of the U.S. Patent Office.

On 21 May 1861, Mortimer Nelson of New York received U.S. Patent no. 32,378 for an aerial car. The craft consisted of an enclosed cabin containing some lifting gas, along with a power plant and crew space. The flat "umbrella" on top of the cabin was actually a plane lifting surface that could be inclined at an angle to the horizon. The unspecified power plant drove twin overhead rotors that could be angled toward the bow or stern to propel the craft forward or backward. Within five years Nelson would abandon the small amount of lifting gas called for in the original patent and opt for a fully heavier-than-air craft. Lift was now supplied by an elliptical disc measuring 35.25 feet long by 10 feet wide. The redesigned craft was powered by a steam engine driving the helicopter blades. In designing a helicopter, the inventor was following the pattern established by A. Masson of Cincinnati. While Nelson's craft was never built, he played a key role in the work of Edward Serrell (Nelson 1866h).

Watson F. Quimby, of Stanton, New Castle County, Delaware, received U.S. Patent no. 33,797 for his "Apparatus for Navigating the Air" on 26 November 1861. Like Nelson, Quimby originally envisioned a helicopter. In subsequent patents issued in 1867, 1872, and 1879, the inventor regressed to ornithopters, or beating-wing machines.

In December 1861 Charles F. Edwards of New York announced that, four years before, presumably in the winter of 1857, he had built and flown a spring-powered aircraft weighing 77 pounds and carrying an extra 66 pounds of ballast. He claimed that the craft, tethered to a central pole with a 30-foot rope, made as many as 117 circles in two minutes. The inventor refused to provide any technical details of his craft but announced plans to build a larger model propelled by a "physico-chemical" power plant. "I need capital," he continued, "and would call the attention of capitalists to this as an investment which promises a splendid return" (Edwards 1861:358).

Arthur Kinsella, of Cascades, Washington Territory, published his 1862 design for an "Improved Flying Machine" in the *Scientific American* (Kinsella 1863). His airship featured a gas bag of woven rattan, covered with silk, and was powered by an engine burning compressed hydrogen and operating two aerial versions of steamboat paddle wheels.

In the fall of 1862, Jeremiah Randall of West Jefferson, Ohio, informed the readers of the *Scientific American* that he had constructed a model aircraft in which a clock spring powered a pair of counter-rotating helicopter blades that could lift his craft a short distance into the air, insisting that "it shows the principle by which a steam engine may be made to travel in the air with or without a balloon" (Randall 1862:198). Randall claimed that he had demonstrated his model to Joseph Sullivan, a leading Ohio scientist of the era, who pronounced it "the first inanimate thing that ever raised itself into the air by its own motive power without a balloon." Sullivan (1862:246) himself offered a disclaimer in the next issue, informing readers that Randall had merely stopped him on the street and described his machine. Sullivan assured readers that the words ascribed to him were a complete fabrication and that he had advised Randall "that he had better save his time and money by directing his attention to something else."

The Union blockade of southern ports inspired William C. Powers, an architect and engineer living in Mobile, Alabama, to develop plans for a steam-powered helicopter capable of dropping bombs that would destroy the Yankee fleet (Figure 9.1). Featuring a boat-shaped hull with a large rudder at the rear, the craft sported two Archimedean screws to lift it into the air and two more screws mounted horizontally along the fuselage to drive it forward. Family legend holds that Powers abandoned the project when a co-worker expressed the fear that the model might fall into enemy hands and be turned against the South.[1]

On 27 April 1863, Dr. R. Findley Hunt, a Richmond dentist, sent President Jefferson Davis a proposal for a steam powered aircraft (Gilman, 1863). Hunt's machine consisted of two or more pairs of wings, a rudder, and a parachute, just in case. Driven by a radial steam engine, the wing pairs stroked up and down. The wing covering, applied in strips, functioned as a valve, opening on the up

Figure 9.1. Helicopter model constructed by William C. Powers of Mobile, Alabama, in 1862. (Photo courtesy of National Air and Space Museum, Washington, D.C., NASM A-34342-A.)

stroke and closing on the down. The operator stood on an open platform just beneath the wings (Hunt 1865).[2] Davis passed Hunt's proposal, along with a set of drawings, through Gen. Robert E. Lee to Col. Jeremy Francis Gilmer, Chief Engineer of the Confederate army, for study and recommendation. Gilmer asked Charles G. Talcott, Superintendent of the Richmond and Danville Railroad, and Lt. Col. A. L. Rives, of the Confederate Army Engineer Corps, to render an opinion. Gilmer forwarded the report to Hunt on 21 July 1863. While Talcott and Rives conceded "the general correctness" of Hunt's scheme, they pointed out that his "estimates and results" were so badly in error that the machine would not fly (Talcott and Rives 1863).

Hunt responded to Gilmer on 28 July, suggesting that "those gentlemen . . . misapprehended to a great extent my plans and views," and admitting that this was "partly my fault in not being more explicit and full in details, and partly my misfortune in not being allowed (as I requested) to make verbal explanations of points that might seem doubtful or objectionable" (Gilmer 1863). Failing to arrange a meeting with the committee or to reverse their decision, he had returned to dentistry, determined to raise the money to build a demonstration model himself. Unable to find a suitable mechanic to assist him, he petitioned both the Secretary of War and Col. Josiah Gorgas, Confederate Chief of Ordnance, to assign Mr. N. Hays, a machinist with Richmond's Tredegar Iron Works, to assist him for 30 days. Failing that, he wrote to President Davis once again on 25 January 1864, explaining that the War Department committee had "condemned my plan without fully understanding my views" and asking him to intervene and arrange the loan of the workman (Hunt 1864).

Hunt may have been a southern patriot, but he was committed to his flying machine. While he had practiced dentistry in Richmond in the mid-1840s, he was living in Washington, D.C., a decade later (Hunt 1845; Hunter, 1853:52). Soon after Lee's surrender he returned to the nation's capital and, on 25 July 1865, petitioned the U.S. Patent Office for a caveat covering his design (Hunt 1865), a preliminary means of protecting his invention until it had been developed to the point at which he could apply for a patent. Two years later he is reputed to have constructed a model of the craft but did not obtain a patent.

In addition to the foregoing projects, at least four flying machine proposals received consideration from either Union or Confederate authorities. In view of the fact that more extensive records exist for these four projects, they deserve special attention.

RICHARD OGLESBY DAVIDSON AND THE ARTIS AVIS

Richard O. Davidson was born in Virginia about 1805, the son of Andrew and Sarah Davidson. In later years he would claim to have become interested in aeronautics in 1839. His first book, *A Disclosure of the Discovery and Description of the Plan of Construction and Mode of Operation of the Aerostat, or a New Mode of Aerostation* (Davidson 1840), was published in St. Louis, where he was apparently living. Throughout his career, Davidson continued his error of referring to his craft as an aerostat, a term referring to a buoyant aircraft. Following a long-winded and not entirely accurate survey of aeronautical history and theory since the work of Roger Bacon in the thirteenth century, the inventor launched into a detailed description of his proposed aircraft, in "the form of an American eagle" (Davidson, 1840:13). Built of whalebone, cork, and rattan, covered with linen and oiled silk, the craft looked like an eagle in every detail, including a beak, coiled spring "legs," and feathers painted on the body, horizontal tail, and wings spanning 24 feet. The "conductor" was housed in the belly of the bird in a cabin measuring four feet on a side furnished with two small windows on the sides and one in front. The main wing spars extended into the cabin and geared to a lever that enabled the operator to flap the wings. In the event that the fellow grew tired, Davidson was willing to consider adding a supplementary carbonic acid gas power plant. (Davidson, 1840:13)

The ornithomorphic "American Eagle" drew an immediate comic reaction from a "member of the L.L.B.B" (Louisville Literary Brass Band), an anonymous Kentucky wag who published a pamphlet entitled *The Great Steam Duck* (Anon. 1841). In this merciless parody of Davidson and all of those who would copy birds too closely, the author suggests a craft shaped like a mallard whose flapping wings would be powered by a steam engine exhausting through the rear of the "bird." The cabin would be furnished with chairs, a table, twin bunks for the crew, a fireplace, poker, tongs, a shovel, candlesticks, and a snuffer. The author assured readers that the only real danger to be encountered by the operator would be experienced as a result of encounters with "sportsmen and others given to the destruction of the feathered race."

Undeterred by his critics, Davidson published a second volume, *A Description of the Aerostat, or a Practical Mode of Aerostation* in New York (Davidson 1841), this time identifying himself as a Virginian, he chose not to repeat his history of aeronautical theory and experimentation, or the "nuts and bolts" level description of "the Aerostat." Instead, he offered an extended justification for his basic design, arguing that "the God of nature" had shaped creatures to navigate in the medium in which they lived. Boat hulls should be generally shaped like fish and flying machines like birds. "No other form of machine nor means of elevating it than those belonging to the birds will at all answer for aerial navigation." Davidson closed his pamphlet with an extended list of the wonderful ends to which his machine could be put and assured readers that "the advantages to be derived from the successful operation of my plan of aerostation, will be as rapid and boundless as the flight of the Aerostat" (Davidson 1841:27).

He published a third volume, *A New Theory of the Flight of Birds*, in Washington, D.C. (Davidson 1858). The 1860 census reported that the 55-year old author was living in Ward Three of the District of Columbia. As the title suggests, it is a study of bird flight. Davidson concludes that the beating action of a bird's wings serves only to raise the bird into the air, while it is the action of gravity and the resistance of the wings to the air that produces forward motion. There is no mention of mechanical flight.

As Confederate leaders were organizing their nation in the early spring of 1861, Davidson travelled from Washington to Montgomery, Alabama, in hope of finding a post in the new government. Failing that, on 10 May 1861 he enlisted in Lowndes County as a private in Capt. Hamilton's Company E, the Prairie Guards, of the 11th Mississippi Infantry. The unit, and presumably Davidson, saw action at First Manassas, then settled into winter quarters at Camp Fisher near Dumfries, Virginia, where, on 14 February 1862, Davidson was discharged with chronic rheumatism. He clearly made an impression on his fellow soldiers, for the discharge was followed by a flurry of notes from officers and men extolling David-son's virtues and recommending him for a clerical post to Postmaster-General John H. Reagan and Secretary of War Judah P. Benjamin. Davidson himself wrote directly to President Jefferson Davis. He apparently served for a time as a stretcher bearer with Capt. John Herbig's Infirmary Company, probably during the Penin-sula Campaign, before returning to civilian life and finding employment first in the Quartermaster General's Office, then as a clerk in the Treasury Department (Davidson n.d.).

Davidson had no doubt about how he could best contribute to a Confederate victory. He later claimed that in 1846 he had offered to build a 50-mile-per-hour aerial craft for the U.S. Post Office and had been rebuffed. At the outset of the war he prepared a memorial "for the aerial locomotion of man," which William S. Barry submitted to the Confederate Congress on 10 May 1861, where it was referred to the Committee on Military Affairs and immediately tabled (CSA 1904–1905, vol. 1:206, 210). On 27 February 1862, Mississippi Sen. Albert G. Brown presented Davidson's petition, "praying for aid in the construction of a machine for aerial navigation," which was once again referred to the Committee on Military Affairs and tabled (CSA 1904–1905, vol. 2:26). Undeterred, David-son presented the petition in person on 14 March 1862, when it was referred to the committee, filed, and forgotten for the third time. In September 1863 he presented his petition to the Virginia legislature, where it was once again ignored. He had no better luck that fall when he submitted his plan directly to the Con-federate War Department (CSA 1904–1905, vol. 2:61).

Davidson presented his proposal to the public in the pages of the *Richmond Daily Dispatch* on 27 January 1864, recounting his attempts to attract official sup-port and promising that contributions of only a dollar apiece would fund the cre-ation of an experimental model of what he now referred to as his "Artis Avis" (Bird of Art), a miracle weapon that would, he assured readers, "rid our country of the perils and privations of this fiendish war" (Davidson 1864a). He proposed the cre-ation of one thousand such craft, each armed with a 50-pound explosive shell and stationed five miles from enemy fortifications. "It will be seen that within a period of twelve hours," he predicted, "one hundred and fifty thousand death dealing bombs could be thus thrown down on the foe, a force that no defensive art . . . could withstand for even a single day; while exposed armies and ships would be instantly destroyed, without the least chance of escape" (Davidson 1864a).

The skeptics were quick to respond. A Mr. T.C.H. (1864) commented at some length, demonstrating "that his scheme must fail, not for want of ingenuity, but because natural laws are fatally opposed." Drawing on his experience with machines and materials, he argued persuasively that an ornithopter of the sort Davidson proposed was impractical. No available power source could match the muscles of a bird. Only if Davidson transformed his design into a buoyant craft, T.C.H. argued, would it leave the ground. Davidson (1864b) responded with a sense of outrage. Clearly, the skeptics misunderstood his plan. Far from constructing a flying machine, the inventor explained that his plan was "less pretentious, more ample and easy of attainment—namely the construction of an artificial bird, which, in traversing the atmosphere, will convey one person from point to point on the earth, and thus accomplish the aerial locomotion of man through the intervention of art."

Many years later Thomas R. Evans (1909:302–303) would recall that an experimenter, surely Davidson, had built a model flying machine in a lumberyard on the east corner of Seventh and Main Streets in Richmond during the war. "There was," he remembered, "an extensive framework composed of rectangular bars of light, white pine." The frame was uncovered, and there was no sign of a power plant or wheels. Davidson told a group of South Carolina veterans that he built and tested such a model. Tied to a flat car and released when the engine was running at top speed, he boasted, "the Bird of Art . . . had no trouble keeping up with the engine without pulling on the rope" (Anon. 1900:304–305). If the inventor's account of such a test flight is not a complete fabrication, it certainly stretches credulity to the breaking point. Thomas Evans (1909:302–303) remembered that the model he saw came to an unhappier end. "One night a strong wind came up. . . . There was a rattling of pine bars . . . and splinters filled the air, and thus fled the hopes of the Confederacy."

At this point, Davidson decided to take his case to those who were closest to the action, the soldiers in the trenches around Petersburg. His first step was to distribute broadsides announcing his coming. "The undersigned," he explained, "is soliciting contributions from the officers and soldiers of the Army of Northern Virginia to construct the . . . [Artis Avis] . . . and to explain how this invention may be employed to destroy or drive from our soil every hostile Yankee, and thus soon close the war" (as quoted in Yee 2008:296). He was not universally welcomed into the camps. "The men were fond of hearing in camp any kind of address, and were an easy prey to sharpers," noted Brig. Gen. Gilbert Moxley Sorrell (1994:94), who had served as Gen. James Longstreet's Chief of Staff before accepting command of a brigade. "He wanted permission to address my men and solicit cash for building his wonderful birds. He was sent out of camp."

Davidson was more successful in other cases. One veteran of McGowan's Brigade of South Carolina Sharpshooters provided some additional details in an account written many years later. On this occasion Davidson seems to have had a model of Artis Avis, "made of hoop-iron and wire." The "Professor" explained that the full-scale craft would be powered by a one-horse power engine "for keeping the wings in motion." Small doors by the shoulder of the "conductor" and at the throat and top of the neck could be opened or closed to cause the "bird" to climb, descend, or turn to the right or left. "In the body of the bird there was room for a number of shells, and the operator, by touching a spring with his foot, could drop them on the enemy from a safe distance" (Anon. 1900:304). The presentation impressed at least some of the Carolinians, a few of whom contributed. The author concluded his account by noting that many years later he encountered a veteran of the 14th South Carolina Volunteers and asked him if he remembered the Artis Avis. "I certainly have heard of it," came the reply, "for I gave a dollar to it" (Anon. 1900:304).

Davidson had visited at least eight brigades of the Army of Northern Virginia camped around Petersburg by March 1865. Reactions varied. One enthusiastic veteran remembered "the intense excitement and joyous hopes pervading the army that the flying bird would exterminate every Yankee in front of Petersburg" (as quoted in Hess 2009:242). Davidson collected $813.50 from one group of brigades. Private Benjamin Simms of the 53rd Georgia informed his sister that he "should as soon look for perpetual motion to be invented as for one of Davidson's birds to rise and fly." The inventor had collected $127 from that brigade, which Pvt. Simms thought "pretty liberal patronage for a humbug" (as quoted in Power 1998:265). What Davidson did with the money he collected is not clear. The Yankees, having narrowly escaped the potential destruction wrought by the Artis Avis, prevailed, and the long aeronautical career of Richard Ogelesby Davidson was at an end.

SOLOMON ANDREWS AND THE AEREON

During the years leading up to the Civil War a number of Americans—Rufus Porter, John Pennington, and Thomas Robjohn, for example—had proposed to establish transcontinental air service using very large cigar-shaped, steam-powered, hydrogen-filled airships. Some of those inventors had established short-lived joint stock companies to fund their schemes and even constructed relatively large flying models of their proposed aircraft. None of them, however, could match the aeronautical career of Solomon Andrews (1806–1872; Figure 9.2).

Born in Herkimer, New York, Andrews spent most of his life in Perth Amboy, New Jersey, where his father, Josiah Bishop Andrews, was minister of the First Presbyterian Church, a physician, health officer of the port, and president of the Middlesex County Medical Society. Solomon was a graduate of the Rutgers Medical School and served as a druggist, physician, health officer, and three-time

Figure 9.2. Portrait of Solomon Andrews. (Photo courtesy of Smithsonian Institution, Washington, D.C., SI-2003-35052.)

mayor of Perth Amboy. He built a national reputation, and his fortune, as an inventor. He developed a tricycle, sewing machine, nutcracker, tobacco filter, barrel-making machine, gas lamp, patent kitchen stove, and a system by which the crew of a moving train could snatch a mail bag without stopping. He once chained $1000 in a chest to a New York lamppost with his patent combination lock and dared the infamous Gotham picklocks to try their hand at retrieving the cash. None was successful.

Andrews had been bitten by the flying bug when, at the age of 17, he found himself daydreaming during one of this father's sermons. "Looking out of the window at the soaring of an eagle in his winding way through the air, I caught as with an electric shock the key to the whole system of aerial flight" (Andrews 1865:4–6). For Andrews, it was the defining moment of his life. "The study of medicine and of the sciences generally, was influenced by one idea. . . . The acquirement of various trades, and skill in workmanship was determined by the resolution to construct a flying machine." Andrews began his aeronautical experiments in the 1820s with a fixed-wing glider, "a paper kite, not attached by any string, but free to move in any direction." With two witnesses present, one of his gliding kites "was made to go some three hundred feet in the open air against a pretty strong wind" (Andrews 1865:4–6). In 1830 Andrews asked the American aeronaut Charles Ferson Durant to test one of his models by dropping it from his balloon. He declined, as did Richard Clayton, another well-known aeronaut.

By the 1840s, Andrews had abandoned the notion of a heavier-than-air machine in favor of a buoyant craft, a navigable airship. He was convinced that experimenters who were struggling to develop an aeronautical power plant were wasting their time and effort. Gravity, he was convinced, could power an airship. "In existing modes of transportation," he argued, "the chief difficulty is to produce motion" (Andrews, 1866:3). Given velocity, "it is easy to give direction and control." He suggested connecting two or more cigar-shaped balloons together. When released, "the difference between the specific gravity of the aerostat, and that of the atmosphere in which it floats, can be used as a propelling power." A spherical balloon, he explained, rose or descended straight up or down, the line of least resistance for a sphere. If made in an elongated form,

> it will ascend or descend in the plane of its longest axis, because it
> meets with less resistance in that direction, and this produces a for-
> ward motion, like sliding on the atmosphere, as a sled slides down a
> hill by the force of gravitation, or as a board rises through the water
> by the same force indirectly applied. (Andrews, 1866:3)

TOM D. CROUCH

In short, the resistance of a suitably shaped envelope to vertical motion could be translated into forward motion, or used to turn the craft, if it could be set at various angles of attack. Half a century later, the pilots of powered lighter-than-air craft would take advantage of this "dynamic lift" to increase performance. Andrews used it to "power" his *Aereon*.

In 1847 Andrews established an Inventor's Institute in Perth Amboy. Within two years he proposed raising $15,000 through the sale of city lots controlled by the institute. The funds would be used to construct a large building that would house "an Aerial Car." He advertised for "two active working men" to be employed in the construction of the Aerial Car and appealed to any potential funders willing to invest in "poor inventors who believe Aerial navigation to be feasible" (Andrews 1847:3). On 21 June 1849 Andrews invited the public to visit the new facility and view the finished frame of the airship, apparently a keel measuring 80 feet long, 20 feet wide, and 10 feet deep. When the exhibition closed, using "a machine invented and constructed by myself," the inventor and his crew cut, sewed, and varnished 1,300 yards of silk into an ellipsoid envelope. The gas bag and frame were connected with cotton netting. A wicker "car" 13 feet long and 20 inches wide was suspended beneath the airship and frame. A "smaller car . . . on runners" was attached to either end of the operator's car and could apparently be moved to the front or rear to control climb or descent. A rudder, 20 feet long, completed the craft. Andrews generated the hydrogen to fill the envelope from steam in five iron retorts, a slow process that required several days to inflate the gas bag. The complete airship was kept inside the building for five weeks while the crew conducted a series of tests, from which "much was learned, and new food for study and reflection was obtained and profited by." While the large craft was never taken outside, the crew did build and release a 12-foot-long aerostat, which "was sent toward the sea and never heard from" (Andrews 1865:7).

The role of Rufus Porter is an especially puzzling element in this early episode in Solomon Andrews' aeronautical career. Porter, publisher of the newly established *Scientific American* and himself an airship experimenter, had apparently been involved with the Inventors' Institute at the outset. By July 1848, however, the journal was condemning both the Aerial Car and its inventor as "hum bugs." Whatever the reason, Andrews would allow this dream of aerial navigation to lie fallow for another 15 years while he grew ever more prosperous (Lipman 1980; Crouch 1983:291–318).

When the Civil War broke out, Andrews immediately volunteered his services as a surgeon and physician, working first with the Sanitary Commission on transports for the sick and wounded and then with the Army of the Potomac stationed at Harrison's Landing on the James River during the Peninsula Campaign of 1862. Observing the work of Thaddeus S. C. Lowe's observation balloons during the fighting along the James, Andrews concluded that "the balloons and balloon corps had cost the government some hundred thousand dollars, and that no information had by their use been derived of the enemy but what was otherwise obtained by scouts and spies." The balloon experiment, he decided, "was a lamentable failure" (Andrews 1865:7).

Recognizing his patriotic duty, Andrews immediately resigned as a surgeon and returned to Perth Amboy to build an airship for the government. Having followed the experience of John Ericsson with the ironclad *Monitor*, he believed that his aerial craft "would not cost one fortieth as much money, and be of far more use to the country." He wrote directly to President Abraham Lincoln on 9 August 1862, offering to build an airship capable of sailing "five or ten miles into Secesia and back again, or no pay." The cost, he noted, would not be much more than one of Lowe's balloons. He would gladly pay that initial cost himself, he added, but for the fact that his capital was tied up in the production of the patent locks for U.S. postal mail bags (Andrews 1865:8).

When he failed to hear from the president within nine days, he wrote to Secretary of War Stanton. This time Peter H. Watson, Assistant Secretary of War, did respond, on 8 September 1862, with a request for a detailed description and drawings of the proposed craft. After delivering the documents in person, and cooling his heels for several days in the offices of the Topographical Engineers, he was informed that while his plan seemed "ingenious in a high degree," the examining officers failed to "perceive how that the invention is of practical utility, and adopted to and needed for the public service." The inventor immediately sent another letter to Stanton (22 September 1862) announcing that he would proceed to build the airship at his own expense and present it to the government "in the hope that it may shorten the war" (Andrews 1865:7–8).

Andrews purchased 1,300 yards of cambric muslin and 1,200 yards of Irish linen and hired John Wise, perhaps the most experienced American aeronaut of the period, to cut, sew, and varnish the linen into 21 spheres, 15 of which were to measure 12 feet in diameter and the remaining 6 spheres 7 feet in diameter. Inflated with hydrogen, these cells would be contained in three cigar-shaped "cylindroids" constructed of the 1,200 yards of Irish linen, which would be held together with a frame and netting to form the lifting portion of what the inventor was once again referring to as an *Aereon* (Figure 9.3). The total cost of construction was less than $10,000. The hydrogen to inflate the craft was generated by mixing dilute sulfuric acid with iron filings in 200-gallon hogsheads at a cost of

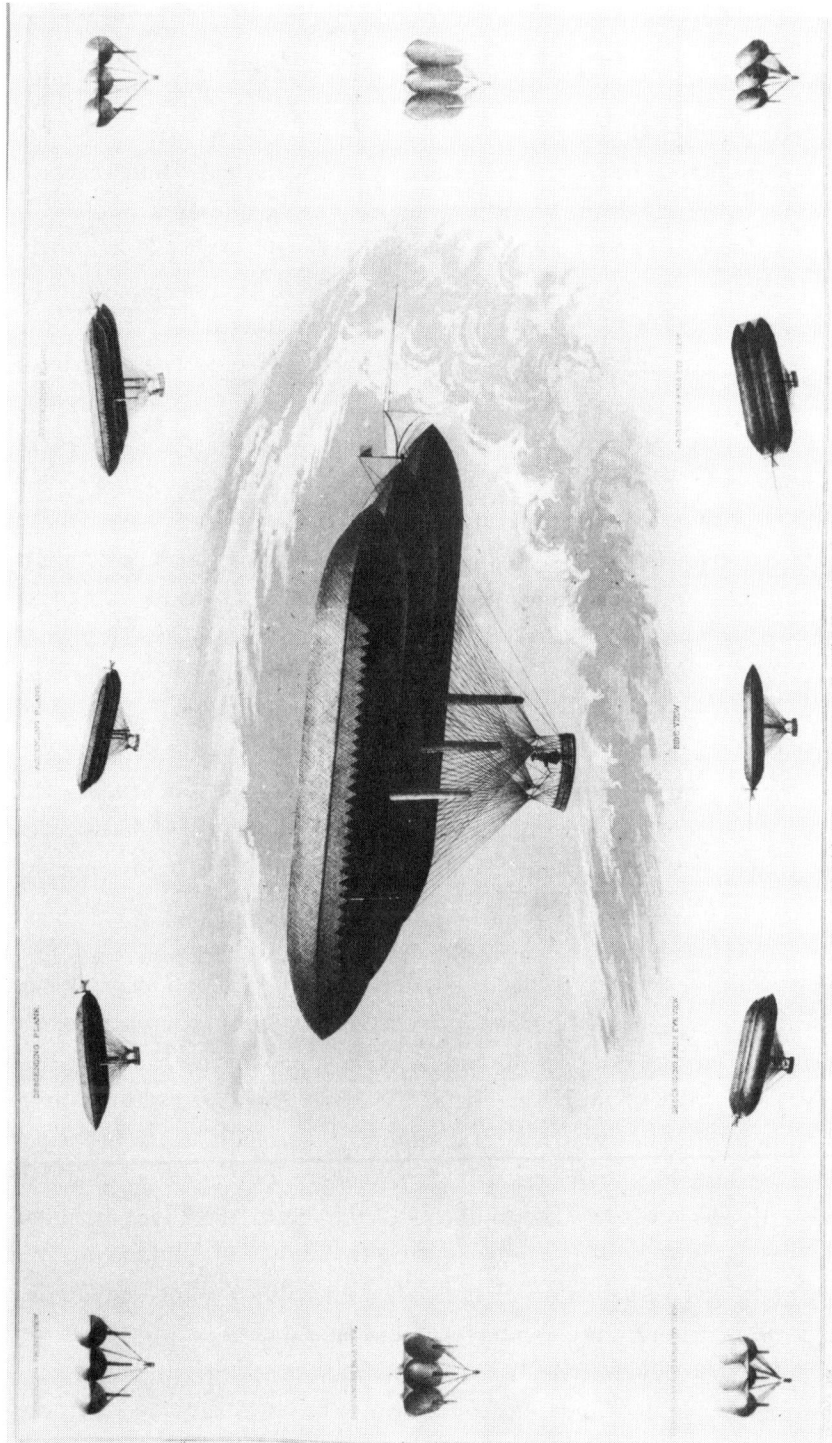

Figure 9.3. The *Aereon* designed and built by Solomon Andrews in 1862–1863. (Print courtesy of National Air and Space Museum, Washington, D.C., NASM-9A05358.)

from $300 to $500, depending on how full the gas bags had to be inflated for a particular trial. "Had there been a gas works in Perth Amboy," the inventor commented, "she could have been filled with carbureted hydrogen gas, each time for about $60" (Andrews 1865:13).

When the individual cells manufactured by Wise collapsed, Andrews removed them and discovered that the varnished outer skin of the cylindroids alone would contain 26,000 cubic feet of hydrogen without excessive leaking. "Its form," the *New York Herald* reported (Anon. 1863), "was that of three cigars pointed at both ends, secured together at their longitudinal equators, covered by a net, and supporting by one hundred and twenty cords a car sixteen feet below, under its center." The operator's car measured 12 feet in length and was 16 inches wide at the bottom. A rudder measuring 17 square feet was provided to steer the craft. The crew began flight testing with the cylindroids alone in July 1863 and with the framework, netting, and rigging in August and September. Andrews wrote to President Lincoln again on 26 August, announcing that his craft was ready for a trial with an operator on board and requesting that a "suitable person" be appointed to witness the trial to confirm that Andrews' airship could perform as advertised. No such official witness was present when Andrews proceeded with flight tests in early September. Fortunately, the *New York Herald*, which had dispatched reporters to Perth Amboy three times that spring and summer to record Andrew's progress, had an observer on hand on 4 September. The *Herald*'s man on the spot offered a somewhat confusing account of events. "The machine made by Mr. Andrews would carry up three men," he explained, "in addition to the fixtures and paraphernalia required for its forward movement. . . . It carried him, weighing 173 pounds, and 256 pounds of ballast" (Anon. 1863; Andrews 1865:12).

The reports of other eyewitnesses give us a better notion of how many flights were made and how often Andrews himself ventured aloft. The Cashier of the City Bank of Perth Amboy, S. V. R. Patterson, described the inventor's June 1863 flight in a letter to President Lincoln dated 8 October 1863: "I saw him go against the wind in his aerial car. He went but a short distance, not ascending over two hundred feet" (as quoted in Andrews 1865:13–14). Ellis C. Waite, an architect and builder, also wrote to President Lincoln on 9 October 1863. He saw Andrews fly three times, "and assisted in getting her out of the building, and to attach the car, which had to be done out of doors, because the ceiling was too low." Andrews "sailed in any direction, either with, by, or against the wind blowing form ten to fifteen miles an hour. He steered her as easily as a sail boat. He went off against the wind, turned her around, and came back to where he started" (quoted in Andrews 1865:13).

Hamilton Fonda, one of Andrew's employees at the U.S. Mail Lock factory, reported to President Lincoln in a letter of 11 October 1863 (as quoted in Andrews 1865:14) that he

> saw the ship inside and out. . . . I helped to build her, and to fill her with gas every time, and aided in all the experiments. . . . I assisted to attach the car in the last trial, on the 4th of September, and to send him off in it. She went upward and forward against the wind, then blowing from the north, not less than ten miles an hour. She minded her helm perfectly, and the rudder was very small for such a large vessel, He turned her around, and came back to the place of starting and came down to the ground. Then he went off to the westward, and turned her head to the east and came down again.

Ex-postmaster John Manning, who saw two of the three flights, informed President Lincoln on 7 October 1863 that Andrews could "navigate her, steer her by a rudder, and come down again. The direction of the wind or air currents seemed to have nothing to do with it" (as quoted in Andrews 1865:14–15). The local constable and tax collector, N. H. Tyrrell, confirmed to President Lincoln in a letter of 10 October that he had seen "the Doctor" fly forward into the wind and maneuver with the rudder (as quoted in Andrews 1865:14–15).

Andrews (1864:1) himself described how he flew the craft:

> To navigate the air in this vessel it is only necessary to step to the rear end of the car, thus elevating the bow five or six degrees, and by throwing out a little ballast she will go ahead on the ascending plane. When she has ascended as high as the aeronaut wishes to go, he opens a valve and discharges some gas, while at the same time stepping toward the forward end of the car, which will depress the bow . . . so . . . she will go ahead on a descending plane. . . . Having forward motion, she is turned just like a boat on the water.

Indeed, Andrews' system would work for a short time. In order to navigate, however, he had to sacrifice either ballast to rise or gas to descend, limiting the distance he could travel and the time he could spend in the air. At the end of his two flights on 4 September, Andrews tied off the rudder and sent the *Aereon* aloft once again with no one on board. The craft rose into the clouds describing great circles a mile and a half in diameter.

She made twenty revolutions before she entered the upper strata of clouds and was lost to view. . . . As she was distinctly seen to move, both below and above the clouds on the clear blue sky at five o'clock p.m., with the sun shining clear on her, there could be no mistake or optical illusion to the beholder. (Anon. 1863:3)

Andrews must have found the *Aereon*, for he reports that "the ship was destroyed by being torn to pieces, the car and valuable fixtures being preserved," to prevent the "secret" of his gravity propulsion system from falling into the hands of the Confederacy. The inventor immediately dispatched letters to "all the governments of Europe." The French responded with a request for drawings and details, while the British simply sent a letter of thanks. He sent a fresh proposal to President Lincoln, along with copies of the letters from those who had witnessed his flights. Those items, he believed, were short-circuited by the War Department before they reached Lincoln's desk.

In December 1863, Senator Ten Eyck and Representative Steele, both of New Jersey, presented a petition to both houses of Congress. Andrews laid out his case, complaining that "nothing but silent contempt has yet been awarded to the most brilliant inventor the world ever saw, by the best Government in the world" (Andrews, 1865:17). When the seven-page petition was initially ignored, Gen. Robert Schenk of Ohio, Chair of the House Military Affairs Committee, re-introduced it. Still, no action was forthcoming. Andrews proceeded to apply for a patent covering the specifics of his system, which was granted (U.S. Patent no. 43,449) on 5 July 1864, and in September 1863 placed the first of three orders with a Paris firm for India rubber models of his *Aereon*. The first two models were unsuccessful. Not until January 1864 was he able to demonstrate a model that would sustain itself in the air. With the assistance of Congressman Schenk, he persuaded Maj. French, the Superintendent of Public Buildings, to allow him to use a room in the basement of the Capitol building to stage a demonstration for the members of the House Military Affairs Committee. "It was a Herculean task," he noted, "to get Senator Wilson, Chairman of the Committee on Military Affairs of the Senate, to witness the spectacle of a flying machine, even in the Capital" (Andrews 1865:23).

The model consisted of inflated cylinders of India rubber, a middle cylinder measuring four feet long by eight inches in diameter, and the two cells on either side just a bit shorter with the same diameter. "These three cylinders were fastened together in the form of a raft," one observer noted, "the middle one being furnished with a light paper plane, which served as a rudder" (Bache et al. 1864:25–27).

The car was suspended under this raft in such a manner that the center of gravity could be shifted each way from the middle toward either end, so that the cylinder might float horizontally, or with the forward end pointing obliquely upwards or obliquely downwards. The vehicle being inflated with gas, and having an extra amount of ascensional power, the car, supplied with bags of sand being so placed that the front end of the aerial vessel will point obliquely upwards, flight was suffered to take place, when, instead of rising vertically, the vessel ascended obliquely in still air along the plane of least resistance in an angle of about 40° to the horizon. . . . When the rudder of the model . . . was slightly turned, the aerial vessel no longer moved in a straight line, but curved around to the right or left . . . as the resistance on one side or the other was increased by the rudder. (Bache et al. 1864:25–27)

The demonstration flight, which occurred sometime in February 1864, was, Andrews noted, "the first time I ever saw a model of my air ship which would sustain its own weight in the atmosphere." The demonstration must have been at least marginally impressive, for the members of the house committee passed a resolution ordering the Secretary of War to appoint "a scientific commission to examine my invention and report to them" (Andrews 1865:24).

On 16 March 1864, Assistant Adjutant Gen. Edward D. Townsend issued War Department Special Order No. 119, appointing Secretary of the Smithsonian Institution Joseph Henry, Alexander Dallas Bache of the U.S. Coast Survey, and Maj. Israel C. Woodruff of the Army Corps of Engineers to a committee to examine Andrews' model and report back to the War Department. Henry and Woodruff met with Andrews for the first time on 21 March and witnessed a demonstration of the model *Aereon* in the library on the west end of the Smithsonian building. "The inventor proved that the balloon can be steered, can be made to move in an oblique direction while ascending or descending," Henry (1862) noted. "The balloon moved horizontally in still air. The power to stem a wind will depend on the amount of ascensional power which the vehicle has to start with." The committee, including Bache, who had missed the first meeting met with Andrews a second time at the Coast Survey office at 7 PM on the evening of 25 March. After the interview with the inventor, the three members retired to the Bache home to discuss the matter and plan a report to the War Department (Henry 1864).

While recognizing Bache as the chair, Henry took responsibility for writing the report, which was dispatched to Secretary of War Stanton on 22 July (Bache et al. 1864). He opened by noting that the officers who had been involved with Thaddeus S. C. Lowe's reconnaissance balloon program earlier in the war had found the results less than fully satisfactory as a result of the low operating altitudes, susceptibility to wind, and oscillations of the balloon, which made it difficult to use binoculars. While noting that Andrews' claims of being able to navigate the air initially seemed "chimerical," discussions with the inventor and a demonstration of a model airship operating on his principle had convinced the members of the committee that "he can really perform what he has asserted he can do" (Andrews 1865:24–25).

The committee had confidence that Andrews could fly a short distance over the enemy in still air and return to a point of safety. Henry admitted that they had only seen a model and did not feel that they could accept the "second hand testimony" of those who had witnessed the full-scale test in New Jersey as evidence in a report to a government department. The group concluded, however, that "there is sufficient possibility of the success of the plan of Dr. Andrews, to warrant . . . an appropriation" that would permit the inventor to construct a second full-scale vehicle for an official test (Bache et al. 1864:26–27).

So near and yet so far. The War Department buried the report, refusing to forward it to Schenk. When the report was delivered, the chairman of the committee found himself powerless to arrange an appropriation. "I stood alone in my Committee as a friend or favorer of the project, willing to make or recommend an appropriation for testing the invention in actual practice," he reported to Andrew on 22 March 1865. "Some of the Committee made distinct and positive opposition; others were willing that I should make a report, but on my own responsibility, and without their support in the House of Representatives, and all but myself were incredulous" (as quoted in Andrews 1865:28).

Having failed to sell his craft as a weapon, Andrews was determined to demonstrate its commercial value. He organized an Aerial Navigation Company on 24 November 1865, with offices at 11 Walker Street, New York City. Thirty-four investors subscribed a total of $6,000, including several large purchases of from $1,000 to $1,500. Andrews valued his own contribution, including his patents and two war surplus balloons at $2,000, raising the total capitalization to $8000. The bylaws left no doubt as to who was in charge, naming Andrews as president and permanent director, as long "as he wants." He nominated both the corporate officers and the board of directors. Those officers were nominated and elected at the first two stockholders meetings, 16 and 25 October 1865. A constitution and bylaws were adopted, and the papers of incorporation filed with the Secretary of State in Albany on 24 November 1865 (Aerial Navigation Co. 1865).

The new firm leased a lot at the corner of Green and Houston Streets in New York for $350 a month and paid an additional $149.62 for grading and $575 for fencing. Andrews had purchased two veteran balloons that had seen service with Lowe's balloon corps—*Intrepid* and *Union*. Workmen cut each balloon into gores, "as one would cut the peel of an orange," and reassembled them into a new cylinder 85 feet long and 50 feet in maximum diameter, *Aereon II*. "It resembled," noted one New York reporter, "a long lemon terminating in a sharp point at either end" (Anon. 1865c). The original control car built for the first *Aereon* 15 years before was suspended beneath the aerostat by means of a complex new arrangement. This time Andrews had fitted a long, wide leather belt along the longitudinal axis and over the top of the aerostat, which could be loosened or tightened by cords in the control car. In this way Andrews hoped to compress the hydrogen in the bag and thus control altitude without sacrificing lifting gas or ballast (Andrews 1866:13–14).

First inflated in April 1866, *Aereon II* took to the air on 27 May, with Andrews accompanied by George W. Trow, C. M. Plumb, and G. Waldo Hill, all officers of the firm, and large stockholders. Following six hours of inflation, the "flying ship" nosed up over the fence. The navigators immediately discovered that the control lines were tangled. They climbed to 200 feet before they could bring their aerial steed under control. Unable to make headway against the wind, they drifted across the East River to a safe landing near Astoria, Long Island (Anon. 1866). If Andrews was less than satisfied with the flight, the citizens of Gotham were delighted. Throngs of gawkers streamed out of the Union Club and the Eclectic, while others craned their necks out of widows or clambered onto roofs. Pickpockets had a field day while pedestrians scrambled to retrieve one of the advertising cards that Andrews had distributed from the sky.

Andrews took to the air again on 5 June, this time accompanied by C. M. Plumb. A reporter for the *New York World* had to step out of the basket when the craft refused to rise. With an enormous crowd watching from inside and outside the fence, *Aereon II* spiraled up while Andrews struggled, once again, to free the rudder control. Ordering Plumb to the back of the car, Andrews dropped ballast and nosed the craft up and out of sight into the clouds. Once again, thousands of distracted New Yorkers looked skyward. "The fair sex seemed . . . oblivious to the fashionable swells," noted one reporter, "while the gentlemen . . . played havoc with the hoops and other various appendages of the perambulating milliner's frames" (Andrews n.d.).

Andrews and Plumb made a safe landing near Oyster Bay, Long Island. The flight had done little to solve the mounting financial problems of the Aerial

Navigation Company, however. The construction of *Aereon II* had drained the treasury. As early as the fifth board of directors meeting on 18 May, treasurer Emmet Dinsmore reported an outstanding debt of $3,100. The two flights had added to the deficit and, in spite of the excitement, had not attracted new investors. With no funds to continue test flight, the company faded away, and with it the long aeronautical career of Solomon Andrews (Aerial Navigation Co. 1865).

EDWARD WELLMAN SERRELL
AND HIS RECONNOITERER

In the spring and summer of 1864, Richard O. Davidson was going from camp to camp passing the hat to fund his Artis Avis. A few miles away, on the other side of the line, Col. Edward W. Serrell (1826–1904), Chief Engineer of the Army of the James, was convincing his commander, Gen. Benjamin Franklin Butler, to send him to Philadelphia, New York City, and New Jersey to build his experimental helicopter, the *Reconnoiterer*, "to be used in the service of this army" (Serrell 1864).

A native of London, England, the 10th of 11 children born to William and Ann Serrell, Edward immigrated to New York City in 1831, where he was educated in city schools. He studied civil engineering under the guidance of his father and an older brother. Like so many nineteenth-century engineers, he would work his way up the profession by moving through a series of ever more responsible positions. In 1845, he became assistant engineer in charge of the Central New Jersey Railroad, moving on to a similar position with other railroads. Three years later he accompanied an expedition that surveyed the route for a railroad from Colon, on the Caribbean coast of Panama, to the Pacific. Returning to the United States in 1848, he married Jane Pound, an English woman, and began work on suspension bridge projects over the Niagara River at Lewiston, New York, and at St. Johns, New Brunswick, Canada. In 1858, he worked as engineer in charge of the Hoosac Tunnel in Massachusetts, the longest tunnel in the United States until 1916, then moved on to the construction of the Bristol Bridge over the England's Avon River. By 1861, he was one of the nation's most experienced civil engineers (Anon. 1888).

In 1860 the U.S. Army Corps of Engineers consisted of just 44 officers and 100 men. With a long war in prospect after the Union defeat at Bull Run, Congress authorized an increase in the number of specialized engineering units. Serrell was authorized by New York authorities to recruit the 1st New York Volunteer Engineer Regiment, which became known as Serrell's Engineers. The unit was accepted by the state on 27 September 1861. Enlisting as a lieutenant colonel, he was promoted to full colonel that December. Ordered to Port Royal, South Carolina, the group became the engineering unit for the 10th Corps. The regiment spent most its time constructing gun emplacements for the siege of Ft. Pulaski. Apparently encountering problems with Chief Engineer Horatio Wright, Serrell spent the months from July to September 1862 under arrest but was acquitted.

He earned special distinction for the design and construction of the Marsh Battery. On 18 July 1863, during the siege of Charleston, South Carolina, Gen. Quincy A. Gillmore ordered Serrell, his chief engineer, to explore the possibility of building a battery for the "Swamp Angel," a 24,000-pound Parrott rifled gun and carriage, in the swampy ground between James and Morris Islands. The gun emplacement was an engineering triumph, combining deep pilings, a log grill, and 13,000 sand bags, 800 tons worth, to create a platform that would literally float on the marsh. The 8-inch Swamp Angel with a 15-pound powder charge fired a 150-pound specially designed incendiary shell, which Gillmore planned to rain down on Charleston rooftops. But the Marsh Battery was a far better example of good engineering than the Swamp Angel was an effective artillery piece. After firing just 35 of the experimental shells, the barrel burst with the 36th (Wise 1994:148).

Serrell spent the spring and summer of 1864 alternating between recruiting duty in New York and service on Gen. Benjamin Butler's staff, functioning at Chief Engineer of the Army of the James during the Bermuda Hundred Campaign and the early months of the Siege of Petersburg. It was during this period that he reconsidered a possibility that had fascinated him for a decade. In 1854 a group of New York and Boston investors asked him to investigate, "in a professional manner . . . the possibility of navigating the air by any process of machinery" (Serrell 1865a).

While considering the matter, Serrell discovered what he described as "a little French toy, which . . . being spun in the hand projected itself into the air upon blades similar to the blades of the propeller of a steamship" (Serrell 1865a). The origins of the little French toy in question are lost in time. Its original designation as a Chinese top may point to its deepest roots. In the West the simplest version of the toy, *un petit moulinet à vent*, or little windmill, first appears in a Flemish manuscript dated to 1325 and reappears in the hands of children portrait sitters during the early modern period. The French naturalist, M. Launoy, and a mechanic, M. Bienvenu, exhibited their version of the helicopter toy before a meeting of the French Academy of Sciences in 1784. For the next century, the toy would inspire one generation after another, from Sir George Cayley, the father of aeronautical engineering, who developed his own version of the hand-launched helicopter at the dawn of the nineteenth century, to 1878, when Bishop Milton

Wright presented 11-year-old Wilbur and his 7-year-old brother Orville with yet another version of the toy. When asked about the roots of their own interest in flight, the brothers always mentioned the gift of the flying toy. While Serrell initially recognized that the little toy offered the germ of an approach to the problem of mechanical flight, he decided, upon reflection, that no available steam engine was capable of providing the required power to lift such a craft off the ground (Crouch 2003:25–26).

Across the Atlantic in 1861, one man, Vicomte Gustave de Ponton D'Amecourt, disagreed, arguing for the possibility of a steam-powered model rotary-wing machine. He stood at the center of a group of like-minded friends, including the pioneer French photographer and balloonist Felix Tournachon ("Nadar"), novelist Jules Verne, and Guillaume Joseph Gabriel de La Landelle, who coined the term *hélicoptère* (from the Greek for twisted wing). In 1863 Landelle assisted D'Amecourt in the design and construction of the first steam-powered model helicopter, which did not fly, and a clockwork version, which apparently did. The group founded *L'Aeronaute* in 1864, the world's first journal dedicated to flight technology. Twenty years later, inspired by their work, Verne armed his antihero of his work, *Robur the Conqueror* (published 1886), with a helicopter (Crouch 2003:38).

There is nothing to indicate that Serrell was aware of developments in France. Over the years since 1854, however, he had paid attention to the increasing power and decreasing weight and bulk of the steam engine, to the point "that the necessary power might be developed from machine so light that it would lift itself into the air, maintain itself in that position and have a sufficient amount still remaining unexpended to propel it at a reasonable velocity." In addition, Serrell noticed an interesting helicopter patent granted to Mortimer Nelson of New York City in May 1861 and visited the inventor to discuss his ideas (Serrell 1865a; see also Nelson 1866:5a).

At the time of the arrival of his unit at Port Royal, Serrell and his engineers were under the command of Gen. Ormsby McKnight Mitchel, a West Point graduate who trained as a lawyer, served as the first engineer of Ohio's Little Miami Railroad, taught mathematics, physics, and astronomy at Cincinnati College, and founded the city's famous observatory. Commissioned a general at the outset of the war, he took command of the Army of the Ohio in the fall of 1861. For his planned attack on Chattanooga, he ordered a team of Union soldiers to steal a locomotive in North Georgia and race north, destroying bridges and track as they approached the city. All the men were captured and some executed. The survivors were awarded the first Medals of Honor.

Recognizing in "Old Stars," as Mitchel was known, a man of vision who might be interested in his notion for a revolutionary aerial weapon, Serrell outlined the idea of a steam-powered helicopter for his commander and demonstrated "a tin model that wound up with a string and a handle . . . and would fly up into the air a hundred feet or more . . . and would carry a bullet or two if the string was pulled hard enough. A version large enough to lift a man, furnished with other propellers to move it forward might carry an observer over enemy lines and back again" (Serrell 1904:952). Impressed, Mitchel ordered Serrell to construct "a machine as would demonstrate the practicality of his method," as soon as he had returned from accompanying a three-day expedition (22–25 October 1862) to Pocotaligo, South Carolina, intended to disrupt the operation of the Charleston and Savannah Railroad. Reporting back to the general, Serrell found Mitchel desperately ill with yellow fever, which soon took his life (Serrell 1865a:2).

During the siege of Charleston, Serrel and fellow engineers, Maj. Richard Butt and Capt. James E. Place, "frequently discussed the details of a machine that should not only take up observers and go where we wished and come back, but carry bombs with high explosives to punish the enemy." Using the atmospheric tables produced by the Ordnance Department, the engineer continued to consider the theoretical possibility of heavier-than-air flight (Serrell 1904:953). In the spring of 1863 Serrell sent his model, drawings, and descriptive material to the War Department, which was acknowledged on 20 April. Just seven days later the colonel received a formal rejection from the Chief of Engineers, via the War Department. "The Engineer Department does not anticipate that the device of Col. Serrell will be free from practical difficulties which have hitherto been found insurmountable in all attempts to rise into and navigate the air, independent of Balloons, and cannot therefore recommend it" (Stanton 1863).

Undeterred, Serrell brought the matter up once again with Gen. Quincy Gillmore, who, on 7 January 1864, ordered him to pursue the matter. However, the transfer of the Union's Xth Corps to Virginia and into action during the Bermuda Hundred campaign west of Richmond and Petersburg intervened. When Gen. Benjamin Butler, commanding the Army of the James, witnessed a demonstration of the model with the four-inch-disc rotor in July 1864, he asked for an additional test. Serrell constructed a six-bladed rotor measuring 12 feet in diameter to determine the lift generated at speeds from 80 to 150 revolutions per minute (rpm). While the blades were set at too great a pitch to allow the rotor to revolve at more than 80 rpm, the results suggested that the general theory might be correct. On 14 November 1864, Gen. Butler asked Gillmore to consider the matter. After reviewing the details of the tests and the drawings Serrell had prepared, Gillmore approved the plan. Butler ordered his chief engineer to proceed to New York and construct a full-scale test craft (Serrell 1865a:2).

The army would not fund the project. Instead, Serrell approached three deep-pocketed petroleum entrepreneurs—Frederick Prentice and Wedworth W. Clark of New York and a Mr. Sully—who responded: "Send the bills to us; we will pay for anything wanted and will help to get it." Ten years after first considering the problem, Serrell was ready to begin work on a flying machine (Serrell 1904:954). The fuselage of the *Reconnoiterer* was to be a copper, cigar-shaped shell measuring 52 feet long, with landing runners on the underside. A chamber at the bottom of the shell would serve as a reservoir for the boiler water, with a second chamber above it for the coal. A light-weight, high-pressure, vertical steam engine with a straight cylinder and vertical boiler were housed in the rear of the shell. The craft would be lifted into the air by twin propellers above and below the fuselage and driven forward by propellers at the front and rear. A large flat copper plate 9 feet across and 45 feet 8 inches long was positioned on either side of the shell. The two plates were connected to a crank running through the fuselage, so that the crew could incline or depress the wings up to six degrees above or below the horizontal. A series of moveable balls were to be used to balance the *Reconnoiterer*. The designer estimated the total weight at take-off with a crew of three on board, along with the required water and fuel for an eight-hour flight, at eight-and-a-half tons (Serrell 1865a:5–8, n.d.).

Work on the fuselage was underway in a workshop in Hoboken, New Jersey. Bennett and Risely, of Greenwich Street, New York City, won the contract for the engine, boiler, gaskets, and everything connected to the propulsion system. As early as 2 January 1865 Serrell reported that the boiler was complete and had been successfully tested to a pressure of 200 pounds. Just six days later, Gen. Ulysses S. Grant removed Gen. Butler from command of the Army of the James and named Gen. Edward O. C. Ord in his place. Serrell had been in the northeast working on his flying machine for only two months when, on 22 January, Ord recalled him to duty to explain the project. Serrell reported on 29 January and was able to convince his new commander that the project was worthwhile.

Back in New York on 10 February, Serrell sent a long and detailed history of the project and description of the *Reconnoiterer*, along with a status report, to Ord. Nine different machine shops and foundries were involved in the project. The shell was complete, as was the engine, which had been tested by pumping water 12 feet into the air. "The shafting," he noted, "is about all turned up. All the small parts are forged. All the small castings are finished. The propellers and elevators are in progress of construction. The supporting frame is being made." The aerodynamic forces operating on the craft had been carefully calculated based on tables in the U.S. Army Ordnance Manual, as well as information from European sources and the records of the Smithsonian Institution (Serrell 1865a:9–10).

Everything seemed to be moving toward a test, but nothing of the sort occurred before the war drew to a close at Appomattox, Virginia, three months later. Work on the project continued as a private venture, however. Since Serrell had raised the money from private investors, the War Department was apparently willing to simply transfer all rights to the inventor. Money continued to flow for a time. There was a bill for $25.93 from a coppersmith on 26 May 1865 and another on 29 May for $514.35 in parts and labor (Bühler 1865; Serrell 1865b).

The project became public knowledge in the summer of 1865. On 1 July, *Scientific American* reprinted an article (Anon. 1865a:1) from the *Journal of Commerce* announcing that "a flying machine is now in process of construction in Hoboken, for the United States Government." The author of the piece traced the project to "the late distinguished General (and Professor) Mitchell" and provided a confused description of the early rotor tests. "The Government toy," the reporter noted, alluding to the little helicopters sold by street vendors, "is a cigar-shaped canoe, built of copper with iron ribs." The engine drove four 20-foot blades, one each on the top and bottom of the "canoe," and one each at the front and rear. The total weight, "fully equipped and manned," was "about six tons" (Anon. 1865a:1). The finished *Reconnoiterer* was apparently a bit smaller and lighter than planned. A waggish critic noted in a subsequent issue that "the thing looks very squally to me" (Anon. 1865b:52).

There is no indication of any attempt to get Serrell's six-ton machine off the ground, but the news stories did draw the attention of Mortimer Nelson, the inventor whom the colonel had visited in the spring of 1861. "Informed that the Government was constructing a Flying Machine in Hoboken," Nelson (1866a:5) wrote,

> I called on the Superintendent of the work, who is the engineer to
> whom I showed my design [in 1861], and learned from him that
> certain experiments had been made in order to test the principle,
> and those experiments proving satisfactory, the building of a car
> had been determined upon, and already commenced.

The two launched a correspondence in the summer of 1866. Like Serrell, who was a prolific inventor, patenting everything from improvements in bridge design and ship's armor to felt hats, silk reels, and folding chairs, Nelson had patented a gold-washing machine, saddles, and improved methods of casting metals. Initially, Nelson was more interested in discussing his own gold-washing machine and Serrell's notion for a new method of driving tunnels than he was in the flying machine problem (Nelson 1866a, b, d).

Nelson had just published his own pamphlet, however, which included a mention of his relationship to Serrell's flying machine effort (Nelson 1866h). By 20 February 1866, he was urging his new friend to join him in a visit to the Novelty Iron Works to see a demonstration of Horatio Allen's new method of harnessing the power of superheated steam, "that is destined to create a revolution in the use of steam power aside from its applicability to the purposes of Aerial Navigation." The plans for his own "aerial car" were developing nicely. "I am in receipt of a communication from Prince Napoleon and also from Mons. De Beaumont, Pres. of the Academy of Sciences at Paris," acknowledging the receipt of his pamphlet and informing him that the Academy had created a committee to study and report on Nelson's ideas (Nelson 1866c).

Nelson now urged Serrell to discuss the creation of a joint stock company to build a *Valamotive*, apparently a new flying machine, presumably to be developed on the basis of Serrell's craft and his own design. He also commented on published accounts of European experiments and the need for Serrell to present the aerodynamic data he had collected during his experiments. Serrell's side of the correspondence with Nelson does not survive (Nelson 1866e, f, g). Serrell had abandoned his project by 1867, never having been able to achieve sufficient steam power to fully operate his machine. He deposited his detailed drawings, the data gathered during his rotor tests, and a "rough model" with the War Department and sent copies of his material to a friend, Mr. Prentice, with offices at 44 Broadway, just two blocks from Serrell's office at 64 Broadway. Finally, he sent a report of his activities to the Duke of Argyll, one of the founders of the Aeronautical Society of Great Britain.

Organized in London in 1866, the Aeronautical Society drew some leading English engineers into membership. Serrell saw the organization as proof that his fellow professionals abroad shared his interest in mechanical flight. As secretary of the society, Frederick William Brearey (1867) responded to his American colleague's report. A great enthusiast for winged flight who was known to launch as many as 20 model gliders into the audience during a lecture, Brearey sent the American a friendly letter with a promise to keep him informed of any developments. Francis Herbert Wenham (n.d.), who would build the first wind tunnel in 1871, also responded with his thoughts on aeronautical propulsion. Serrell must have been especially pleased to receive a response from Sir William Charles

Fairbairn (1867), 1st Baronet of Ardwick. One of the world's leading civil engineers, Fairbairn had designed and overseen the construction of the famous Britannia Bridge and pioneered the development of both steamships and locomotives in Great Britain. He admitted to membership in the new organization and to an interest in the atmosphere but was also quite skeptical as to the future of winged flight. "As to navigating the air," he remarked to Serrell,

> there does not appear to me, in the present state of knowledge,
> to be any chance of success, and altho' some of the members are
> more sanguine than myself, I must nevertheless look upon all our
> attempts in that direction as . . . Utopian.

Should, however, he become aware of any advances in the field, Fairbairn promised to pass them along to his American colleague.

Serrell was ready to agree. "My conclusion," he remarked many years later, "was that at that time no existing machine could develop power enough to fly mechanically, without the use of gas-holders" (Serrell 1904:952). He never lost his interest in flight, however, or his faith that one day it would be accomplished. In 1904, 40 years after he had begun work on his *Reconnoiterer*, he noted news reports of the first flights of the Wright brothers at Kitty Hawk, North Carolina. The first erroneous accounts of those flights suggested that the brothers had used a helicopter blade to lift their machine into the air, leading Serrell (1904:955) to suppose that "their machine is very much like, if not identical with, the army machine [of 1864–1865]." In any case, he concluded, "they are to be most heartily congratulated upon the measure of success that has crowned their efforts, and this kind thought extends to my friend of years gone by—[Octave] Chanute—who is reported to have helped them." If Serrell could not be credited with success, at least a fellow civil engineer of his generation, Octave Chanute, had played a role in the achievement of the final goal.

CONCLUSION

What are we to make of this generation of Americans who caught the flying machine bug half a century early? One thing is clear, while most of their practical, down-to-earth fellow citizens may have lumped anyone who dreamed of heavier-than-air flight in with cranks, crooks, and mountebanks, that opinion was not universal. *Scientific American* was a serious journal with a wide national circulation. The large number of articles on aspects of the subject appearing before, during, and after the Civil War, coupled with the significant number of patents issued, indicates that mechanical flight was of at least casual interest to a significant number of Americans with a general interest in technology.

The rise in the appearance of flying machine schemes during the closing phase of the Civil War is surely no accident. In his account of the prehistory of the submarine, historian Alex Roland (1977) argued that visionary technological solutions to intractable strategic problems often seem more appealing to a nation facing military defeat at the hands of a larger and more powerful foe. Whether to the Confederate troops suffering in the trenches at Petersburg in 1864, or the Nazi high command in 1944, the siren call of a wonder weapon capable of miraculously reversing the tide may seem very appealing.

But the would-be aviators of the mid-nineteenth century were pursuing a dream for which the technological preconditions were not yet in place. From a practical knowledge of wing design and aerodynamics to the development of a light-weight power plant, the missing bits and pieces of knowledge and technology would become available over the next three decades. Lewis Mumford (1934:23–24) had something to say about technological dreamers in such a situation.

> Looking at the birds, men dreamed of flight, perhaps one of the
> most universal of man's envies and desires. The dream gives direc-
> tion to human activity and both expresses the inner urge of the
> organism and conjures up appropriate goals. But when the dream
> strides too far ahead of fact, it tends to short-circuit action: the
> anticipatory subjective pleasure serves as a surrogate for the thought
> and contrivance and action that might give it a foothold in reality.
> The disembodied desire, unconnected with the conditions of its
> fulfillment or with its means of expression, leads nowhere.

Well, not quite nowhere. The decade of the Civil War represented a turning point in the history of flight in America and the world. Before that time, the field had indeed been dominated by folk technicians whose ideas would not materially contribute to the goal of powered winged flight. The oldest flying machine dreamers were captives of what Mumford identified as the "obstacle of animism." From Leonardo to Richard O. Davidson, they drew direct technological inspiration from the beating wings of birds. The trick was to re-imagine the basic principles of bird flight in a more efficient, mechanical form. That would ultimately require a separation of lift and propulsion into distinct systems and a merger of the secrets of the wing with the basic principles of bridge design, an imaginative leap quite beyond the apparent capabilities of Richard O. Davidson.

Solomon Andrews, who employed aerostatic principles to get off the ground and aerodynamic ideas to navigate his craft, was at the opposite end of the spectrum. He applied his own twist to well-understood technology. The earliest successful airship designers, beginning with Henry Giffard in 1852, had employed steam or electrical power to propel their craft at speeds of less than 15–20 miles per hour. Such an airship could not be operated in even a light wind. With no better power plant available, Andrews applied the principle of dynamic lift, relying on the resistance of the air to the vertical movement of the envelope to provide a very limited degree of maneuverability. Obviously, the need to valve gas and drop ballast to initiate the vertical movement required to maneuver would limit the amount of time the craft could remain in the air. Still, Andrews was the only one of our Civil War dreamers to get off the ground and the only one whose craft, though scarcely practical, might actually have been able to fly a very short distance over enemy lines and return—under perfect conditions.

While Edward Serrell would never get off the ground, neither does he deserve to be dismissed as a hopeless dreamer. He represented a new generation of aeronautical experimenters drawing on the experience of contemporary engineering practice and operating much closer to what we might identify, with benefit of whiggish hindsight, as the "mainstream" of evolving flight technology. That tradition was rooted in the work of Sir George Cayley (1773–1857), a Yorkshire baronet who was the first to reach back into the embryonic engineering literature to identify the work of a handful of men like John Smeaton, whose research and ideas would be useful in aeronautics. He conducted his own series of well-considered tests, designed and built the world's first successful model airplane (1804), developed a wide range of configurations, and built and flew at least three gliders capable of carrying human beings into the air.

The A. Masson of the Cincinnati "Aerial Steamboat" of 1834 is clearly in the Cayley tradition. Proof of that came in 1843, 10 years after Masson unveiled his craft, when Cayley himself designed a helicopter machine very similar to Masson's. The Cincinnatian was working in what was for that time the aeronautical mainstream, drawing on promising design ideas that were inspiring other leading experimenters in other parts of the world.

The little helicopter toy whose history is briefly traced above was a universal aeronautical constant. From early modern times, the only man-made objects capable of sustained flight for at least short periods were the kite and the little toy helicopter. From the time of Cayley to that of the Wright brothers, the kite inspired the design of fixed-wing aircraft and provided a method of testing flight structures, aerodynamic principles, and flight control. As noted above, the little helicopter inspired flying machine experimenters from the beginning of the nineteenth century to the end. Serious engineers like Serrell, who favored that design, were thus also in the aeronautical mainstream, drawing from a relatively rich store of ideas shared by experimenters on both sides of the Atlantic.

Professional engineers entered the field in a serious way in the 1860s. They founded the world's first aeronautical societies to draw fellow professionals into membership. The Aeronautical Exhibition of 1868, sponsored by the Aeronautical Society of Great Britain in London's Crystal Palace, suggested to the general public that aviation was a matter of interest and study to serious technicians. They published the first journals covering flight technology and established research programs leading to the creation of such key engineering tools as the wind tunnel. Serrell's work, grounded on his rotary-wing tests, is proof that this movement toward the professionalization of aeronautics had crossed the Atlantic.

And Serrell was not alone. Mechanical flight was the subject of the very first presidential address offered to the newly organized American Society of Mechanical Engineers. Speaking in 1881, President Robert Henry Thurston (as quoted in Reeve 1953:988), professor of engineering at the Stevens Institute of Technology, suggested that there was "no department of engineering in which the art of the mechanic has opportunity for greater achievement." While balloons and airships were all fine and well, he argued that the "real promise lies in the direction of flying machines lifted by their own power, not buoyed up by gas." Thurston concluded by suggesting that someone in his audience might "in our own day win the fame that awaits the first successful builder of a flying machine." As Thurston spoke, 14-year-old Wilbur Wright and his 10-year-old brother Orville may have still been building and flying their own versions of the little helicopter toy their father had given them three years before.

Notes

1. Correspondence contained in Accession File 156895, National Air and Space Museum, Washington, D.C. Presented to the museum by Powers' descendants in 1940, the original model was destroyed in an accident at a museum storage facility and de-accessioned. Photos of the model and mechanical drawings associated with the project are held by the Archive of the National Air and Space Museum.

2. A scrapbook of Hunt Papers was auctioned to an anonymous bidder in 2011. All documents cited relating to Hunt were printed from scans of items in that collection provided by Mr. Bobby Livingston of the RR Auction House. All documents cited with permission of the RR Auction House. The document in question was presented to the U.S. Patent Office in the months after the war. It is apparent, however, that it describes the machine originally presented to Confederate authorities.

References

Abbreviations

AIAA   American Institute of Aeronautics and Astronautics

NARA   National Archive and Record Administration

RG   Record Group

Aerial Navigation Co. 1865. "The Minute Books and Certificate of Incorporation of the Aerial Navigation Company, Dr. Solomon Andrews, President, New York, 1865." Box 174, AIAA Collection, Manuscript Division, Library of Congress, Washington, D.C.

Andrews, Solomon. 1847. *Inventors' Institute.* Perth Amboy, N.J.: n.p.

———. 1864. Aerostat. U.S. Patent no. 43,449, 5 July.

———. 1865. *The Art of Flying.* New York: John F. Trow.

———. 1866. *The Aereon, or Flying Ship.* New York: John F. Trow.

———. (n.d.) Undated news clippings. AIAA Historical Scrapbooks, Manuscript Division, Library of Congress, Washington, D.C.

Anonymous. 1822a. "Seventeenth Congress—first session, House of Representatives, Monday, March 25," *Niles' Register,* 30 March, 79.

———. 1822b. "Seventeenth Congress—first session, House of Representatives, Monday, March 25," *Niles' Register,* 6 April, 95.

———. 1823. "On Steam Aerostation," *Daily National Intelligencer,* 17 November, 2.

———. 1824. "Ballooning." *Mechanic's Magazine,* 2 (12 Jun.), 222. [London.]

———. 1834a. *Cincinnati Directory.* Cincinnati, Ohio: E. Deming, 1834:112.

———. 1834b. "Novelty Aerial Steamboat." *Liberty Hall and Cincinnati Gazette,* 3 July, 4.

———. 1834c. Aerial Steamboat. *American Railroad Journal and Advocate of Internal Improvements,* 16 August:502.

———. 1834d. "Flying Machine." *Daily Cincinnati Republican,* 23 August, 2.

———. 1834e. "To the Public—Flying Machine." *Daily Cincinnati Republican,* 22 October, 2.

———. 1834f. "Flying Machine." *Cincinnati Chronicle and Literary Gazette,* 25 October, 2.

———. 1841. *The Great Steam Duck; or, A Concise Description of a Most Useful and Extraordinary Invention for Aerial Navigation.* Louisville, Ky.: Henkle, Logan.

———. 1860a. $1,000 Reward—A Flying Machine Wanted. *Scientific American,* (4 Aug.):88.

———. 1860b. Aerial Navigation. *Scientific American,* (18 Aug.):116.

———. 1860c. Flying Machines in the Future. *Scientific American,* (8 Sept.):165.

———. 1863. "Aerial Navigation." *New York Herald,* as reprinted in the *Chicago Tribune,* 11 September, 3.

———. 1864a. Can We Fly by Steam. *Scientific American,* (24 Aug.):137.

———. 1864b. Flying Impossible. *Scientific American,* (27 Aug.):135.

———. 1864c. Pneumatics. *Scientific American,* (27 Aug.):135.

———. 1864d. The True Plan for a Flying Machine. *Scientific American,* (15 Oct.):246.

———. 1865a. A Steam Flying Machine. *Scientific American*, (1 July):1.

———. 1865b. The Government Flying Machine. *Scientific American*, (22 July):52.

———. 1865c. Undated news article. In Aerial Navigation Co. 1865.

———. 1866. "Aerial Navigation," *New York Commercial Advertiser*, reprinted in the *Chicago Tribune*, 1 June, 2.

———. 1888. "Edward W. Serrell." In *Appleton's Cyclopedia of American Biography*, ed. James Grant Wilson and John Fiske, p. 464. New York: D. Appleton.

———. 1900. A Confederate Airship. *Southern Historical Society Papers*, 28:304–305.

Bache, Alexander Dallas, Joseph Henry, and Israel C. Woodruff. 1864. Report of the Scientific Commission to Edwin M. Stanton, 22 July. Reproduced in Andrews (1865), pp. 25–27.

Brearey, Frederick William. 1867. Letter to Edward Wellman Serrell, 25 May. Serrell Papers.

Bühler, C.F. 1865. Bill from copper-smith, 27 May. Serrell Papers.

Crouch, Tom. 1983. *Eagle Aloft: Two Centuries of the Balloon in America*. Washington, D.C.: Smithsonian Institution Press.

———. 2003. *Wings: A History of Aviation from Kites to the Space Age*. New York: W.W. Norton.

CSA (Confederate States of America). 1904–1905. *Journal of the Congress of the Confederate States of America, 1861–1865*. 7 vols. Washington, D.C.: Government Printing Office. http://memory.loc.gov/ammem/amlaw/lwcc.html (accessed 2 May 2013).

Davidson, Richard Oglesby. 1840. *A Disclosure of the Discovery and Description of the Plan of Construction and Mode of Operation of the Aerostat; or, A New Mode of Aerostation*. St. Louis, Mo.: n.p.

———. 1841. *A Description of the Aerostat; or, A Practical Mode of Aerostation*. New York: William Applegate.

———. 1858. *A New Theory of the Flight of Birds*. Washington, D.C.: Henry Polkinhorn.

———. 1864a. "To the Officers, Soldiers, and Citizens of the Confederate States." *Richmond Daily Dispatch*, 27 January 1864.

———. 1864b. "Flying Machines." *Richmond Whig*, 22 March 1864.

———. (n.d.) Compiled service record, 11th Mississippi Infantry. NARA RG 309, Box Bu-D, Microfilm Publication 232, roll 192, Washington, D.C.

Edwards, Charles F. 1861. Suggestions about Flying—The Thing Accomplished. *Scientific American*, (7 Dec.):358.

Evans, Thomas R. 1909. Tells Story of Flying Machine of Confederacy. *Southern Historical Papers*, 37:302–303.

Fairbairn, William Charles. 1867. Letter to Edward Wellman Serrell, 8 March. Serrell Papers.

Gilmer, Jeremy Francis. 1863. Letters from Gilmer to R.F. Hunt, 21 July 1863 and 30 July 1863. Author's collection. [Refers to Hunt's first letter to Jefferson Davis.]

Henry, Joseph. 1862. J. Henry's locked book entry, 22 March. Joseph Henry Papers Project, item SIA2012-3313, Smithsonian Institution Archives, Washington, D.C.

———. 1864. Letter to A.D. Bache, 25 March. Alexander Dallas Bache Papers, RG 7053, box 4, Smithsonian Institution Archives, Washington, D.C.

Hess, Earl J. 2009. *In the Trenches at Petersburg: Field Fortifications & Confederate Defeat*. Chapel Hill: University of North Carolina Press.

Hunt, R. Findley. 1845. Dental bill, August. M74SS IT211D3723, Tayloe Family Papers, Southern History Collection, Archives and Special Collections, University of North Carolina Library, Chapel Hill.

———. 1864. Letter to J. Davis, 25 January. Author's collection.

———. 1865. Petition to U.S. Patent Office, 25 July. Author's collection.

Hunter, Alfred, ed. 1853. *The Washington and Georgetown Directory*. Washington, D.C.: Kirkwood and McGill.

Jefferson, Thomas. 1822. Letter "From Thomas Jefferson to D.B. Lee, 27 April 1822," Founders Online, National Archives (http://founders.archives.gov/documents/Jefferson/98-01-02-2785 [last update: 2015-09-29]; accessed 20 November 2015).

Johnston, David E. 1980. *The Story of a Confederate Boy in the Civil War*. Radford, Va.: Commonwealth Press. [Originally published Portland, Or.: Glass and Prudhomme Co., 1914.]

J.T.D. 1864. Aerial Navigation. *Scientific American*, (17 Sept.):182.

Kinsella, Arthur. 1863. Improved Flying Machine. *Scientific American* (29 Aug.):129.

Lipman, Jean. 1980. *Rufus Porter Re-Discovered: Artist, Inventor, Journalist, 1792–1884*. New York: Clarkson Potter.

Miller, Perry. 1979. "The Responsibility of Mind in a Civilization of Machines." [Reprint.] In *The Responsibility of Mind in a Civilization of Machines*, ed. John Crowell and Stanford J. Searl Jr., pp. 195–213. Amherst: University of Massachusetts Press. [Originally published in *American Scholar*, 31, no. 1(Winter 1961–1962):51–69, 1961–1962.]

Mumford, Lewis. 1934. *Technics and Civilization*. New York: Harcourt Brace.

Nelson, Mortimer. 1866a. Letter to Edward Wellman Serrell, 12 January. Serrell Papers.

———. 1866b. Letter to Edward Wellman Serrell, 25 January. Serrell Papers.

———. 1866c. Letter to Edward Wellman Serrell, 20 February. Serrell Papers.

———. 1866d. Letter to Edward Wellman Serrell, 26 March. Serrell Papers.

———. 1866e. Letter to Edward Wellman Serrell, 6 July. Serrell Papers.

———. 1866f. Letter to Edward Wellman Serrell, 9 July. Serrell Papers.

———. 1866g. Letter to Edward Wellman Serrell, 4 August. Serrell Papers.

———. 1866h. *Mortimer Nelson's Aerial Car*. New York: n.p.

Pennington, John H. 1838. *Aerostation; or, Steam Aerial Navigation*. Baltimore: n.p.

Porter, Rufus. 1860. Flying Machines in the Future. *Scientific American*, (8 Sept.):125.

Power, J. Tracy. 1998. *Lee's Miserables: Life in the Army of Northern Virginian from the Wilderness to Appomattox*. Chapel Hill: University of North Carolina Press.

Randall, Jeremiah. 1862. A Practical Flying Machine. *Scientific American*, (27 Sept.):198.

Reeve, Harrison F. 1953. AME Role in Powered Flight. *Mechanical Engineering*, (Dec.):987–999.

Riley, Franklin Lafayette. 1988. *Grandfather's Journal*. Ed. Austin C. Dobbins. Dayton, Ohio: Morningside Press.

Roland, Alex. 1977. *Underwater Warfare in the Age of Sail.* Bloomington: Indiana University Press.

Serrell, Edward Wellman. (n.d.) Undated note. Serrell Papers.

———. 1864. Return, 11th Engineers of N.Y., Col. E.W. Serrell. 12–30 April 1862, "Oct. 1864." 8. RG 94, M 2004, NARA Cat. no. 300398, NARA, Washington, D.C.

Serrell, Edward Wellman. 1865a. Letter to General [Ord], 10 February. Serrell Papers.

———. 1865b. Letter to Wight and Co., 29 May. Serrell Papers.

———. 1904. A Flying Machine in the Army. *Science*, new series, 19, no. 493(24 June):952–955.

Serrell Papers. 1864–1867. Edward Wellman Serrell Papers. Collection 2011-0040, National Air and Space Museum Archives, Washington, D.C.

Sorrell, G. Moxley. 1994. *Recollections of a Confederate Staff Officer.* New York: Smithmark. [Reprint. Original publication New York and Washington, D.C.: Neale, 1905.]

Stanton, Edward M. 1863. Letters sent by the Secretary of War relating to military affairs, 9 March, 17 August. Letters Sent by the Secretary of War Relating to Military Affairs, 1800–1889, vol. 52–52A, 9 March–17 August 1863, M6 Roll 51, NARA, Washington, D.C.

Sullivan, Joseph. 1862. The Flying Machine—A Disclaimer. *Scientific American*, (16 Oct.):246.

Talcott, Charles G., and A.G. Rives. 1863. Letter to R. Findley Hunt, 21 July. Author's collection.

T.C.H. 1864. "Flying Machines." *Richmond Whig*, 19 March 1864.

Tocqueville, Alexis de. 2003. "Why Americans Are More Addicted to Practical Rather Than Theoretical Science." In *Democracy in America*, trans. Henry Reeve, vol. 2, chap. 10, pp. 41–47. New York: Barnes and Noble. [Originally published New York: J. and H.G. Langley, 1840, 2 vols.]

Wenham, Francis Herbert. (n.d.) Letter to Edward Wellman Serrell, undated. Serrell Papers.

Wise, Stephen R. 1994. *Gate of Hell, Campaign for Charleston Harbour, 1863.* Columbia: University of South Carolina Press.

Yee, Gary. 2008. Champion of Confederate Airpower or Charlatan. *Military Collector and Historian,* 60, no. 4(Winter):296–299.

ABOUT THE CONTRIBUTORS

Tom D. Crouch, Ph.D. (Ohio State University, 1976), is senior curator of aeronautics at the Smithsonian Institution's National Air and Space Museum. A Smithsonian employee for more than four decades, he is a prize-winning author/editor of 15 books and numerous articles for both scholarly and popular journals. He is a fellow of the American Institute of Aeronautics and Astronautics and, as a presidential appointee, chaired the advisory board to the federal Centennial of Flight Commission.

David J. Gerleman, Ph.D. (Southern Illinois University, 1999), is an assistant editor with the Papers of Abraham Lincoln, a long-term project dedicated to identifying, imaging, and publishing all documents written by or to Abraham Lincoln. He also teaches courses on nineteenth-century America, the Civil War, and Abraham Lincoln at George Mason University. He resides in Alexandria, Virginia.

Seymour E. Goodman, Ph.D. (California Institute of Technology, 1970), is Professor of International Affairs and Computing and Director Emeritus of the Sam Nunn Security Program at Georgia Institute of Technology. His historical interests include the study of the roles of technology in large-scale conflicts, particularly the American Civil War, World War II, and the Cold War, and conflict in cyberspace.

Barton C. Hacker, Ph.D. (University of Chicago, 1968), is senior curator of armed forces history in the Smithsonian Institution's National Museum of American History. He is also the founding editor of *Vulcan: The International Journal of the Social History of Military Technology.* He has lectured widely and published on the histories and historiographies of military technology, non-Western military institutions, women and armies, nuclear weapons testing, and manned spaceflight. In 2003 he received the Leonardo da Vinci Medal of the Society for the History of Technology.

John A. Macaulay, Ph.D. (University of South Carolina, 1998), is the author of *Unitarianism in the Antebellum South: The Other Invisible Institution* (2001).

Formerly a research associate with the Papers of Abraham Lincoln project at the Library of Congress and professor at Erskine College, S.C., he is now a senior training and communications specialist at the State Department.

Merritt Roe Smith, Ph.D. (Pennsylvania State University, 1971), is the Leverett and William Cutten Professor of the History of Technology at MIT. His primary research and teaching interest is American industrialization, particularly the role of the military as a catalyst of technological change. He is the prize-winning author or editor of seven books and many articles. Among his numerous awards and honors is the Leonardo da Vinci Medal of the Society for the History of Technology.

Steven A. Walton, Ph.D. (University of Toronto, 1999), has a background in both mechanical engineering and the history of science and technology. He teaches in Michigan Technological University's Industrial Archaeology program. His studies of arms and armor, gunpowder artillery, fortification, torpedoes, and scientific instruments all pivot around his interest in how the military, industry, and ideas come together to produce weaponry.

Sarah Jones Weicksel is a graduate of the Winterthur Program in American Material Culture and a doctoral candidate in history at the University of Chicago. As a 2015–2016 Committee on Institutional Cooperation–Smithsonian Institution Fellow at the National Museum of American History and the National Museum of African American History and Culture, she is completing a dissertation on the politics of clothing production, consumption, and destruction in the American Civil War era.

Jorit Wintjes, Ph.D. (Julius-Maximilians-Universität Würzburg, 2003), Dr. phil. habil. (2013), currently works as a senior lecturer for ancient history at Julius-Maximilians-Universität Würzburg. His main research interests include ancient history and nineteenth-century naval history. Among his recent publications are articles on Roman naval affairs, Caesar's latter-day military reputation, women in ancient armies, Roman field artillery, and Confederate torpedo boats.

INDEX

01 14